Laboratory Manual of Biomathematics

Laboratory Manual of Biomathematics

Raina S. Robeva, James R. Kirkwood,
Robin L. Davies, Leon S. Farhy, Michael L. Johnson,
Boris P. Kovatchev, and Marty Straume

Sweet Briar College
Sweet Briar, VA

University of Virginia
Charlottesville, VA

AMSTERDAM • BOSTON • HEIDELBERG • LONDON
NEW YORK • OXFORD • PARIS • SAN DIEGO
SAN FRANCISCO • SINGAPORE • SYDNEY • TOKYO
Academic Press is an imprint of Elsevier

Academic Press is an imprint of Elsevier
30 Corporate Drive, Suite 400, Burlington, MA 01803, USA
525 B Street, Suite 1900, San Diego, California 92101-4495, USA
84 Theobald's Road, London WC1X 8RR, UK

This book is printed on acid-free paper. ∞

ISBN: 978-0-12-374022-9

For all information on all Elsevier Academic Press publications
visit our Web site at www.books.elsevier.com

Printed and bound in the United Kingdom
Transferred to Digital Printing, 2010

CONTENTS

PREFACE

This laboratory manual is a collection of laboratory projects that are related to chapters in the text *An Invitation to Biomathematics*. Although all projects are related to chapters in the textbook, where the concepts are described in detail, each of the laboratories contains a brief introduction to the necessary biological, mathematical, and computer software backgrounds. Thus, the use of the text *An Invitation to Biomathematics* is recommended but not required. Each lab project was designed to encourage hands-on exploration with an emphasis on developing and validating mathematical models. For students who will be using both the text and laboratory manual, the lab projects and their corresponding text chapters are as follows: Labs 1, 2, and 3 (Chapter 1); Labs 4 and 5 (Chapter 2); Lab 6 (Chapter 3); Lab 7 (Chapter 4); Lab 8 (Chapter 5); Lab 9 (Chapter 6), Lab 10 (Chapter 9); Lab 11 (Chapter 10); and Lab 12 (Chapter 11).

An important feature of this collection is that the projects, except for the introductory Lab 1, are largely independent from one another. They can be used as a set of laboratories in a designated mathematical biology course, but they can also be used independently as stand-alone modules in conventional mathematics and biology courses, as presented in the following table.

Project	Biology Courses	Mathematics Courses
Lab 1	Ecology	Precalculus, Discrete Dynamical Systems, Calculus, Ordinary Differential Equations, Mathematical Modeling
Lab 2	Ecology	Precalculus, Calculus, Ordinary Differential Equations, Mathematical Modeling
Lab 3	Pharmacology	Precalculus, Discrete Dynamical Systems, Mathematical Modeling
Lab 4	Microbiology, Epidemiology	Calculus, Ordinary Differential Equations, Mathematical Modeling
Lab 5	Ecology	Calculus, Ordinary Differential Equations, Mathematical Modeling
Lab 6	Genetics	Precalculus, Discrete Dynamical Systems
Lab 7	Genetics	Probability, Statistics
Lab 8	Endocrinology	Statistics, Mathematical Modeling
Lab 9	Physiology	Statistics, Mathematical Modeling
Lab 10	Endocrinology	Calculus, Mathematical Modeling
Lab 11	Endocrinology	Ordinary Differential Equations, Mathematical Modeling
Lab 12	Cell Biology, Physiology	Mathematical Modeling

Software for completing the lab projects is a follows (note: laboratory projects requiring software are listed in parentheses):

Software for solving difference equations and determining numerical solutions for ordinary differential equations. The specific syntax used throughout the laboratory manual is that of *BERKELEY MADONNA*, although any similar product, such as VENSIM or STELLA, can be used. We chose *BERKELEY MADONNA*; a fully functional version (with some limitations on saving and printing) can be downloaded free of charge from www.berkeleymadonna.com. (Labs 1, 2, 3, 4, 5, 6, and 11.)

Software for statistical data analysis. The specific syntax we use is that of MINITAB, but any comparable system for statistical analysis (e.g., SPSS) can be used. (Labs 7 and 9.)

A spreadsheet application. We chose Microsoft Excel, because it is readily available on most computers. (Labs 8 and 9.)

Research software Pulse_XP. Available for download free of charge from http://mljohnson.pharm.virginia.edu/home.html. (Lab 10.)

Lab 12 is based on computer software output from a variety of research software routines. The outputs are provided with the project and, therefore, students do not need to use any software on their own.

All projects have been tested in the interdisciplinary course Topics in Biomathematics and in conventional courses in Genetics, Microbiology, Calculus, Statistics, Probability, and Mathematical Modeling, at Sweet Briar College. Selected projects have also been tested at Bates College, the University of Virginia, and the University of Massachusetts, Boston.

The Authors
July 20, 2007

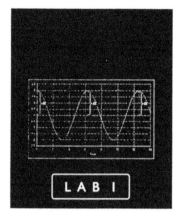

Exploring *BERKELEY MADONNA* in the Context of Single-Species Population Dynamics

This laboratory exercise serves as an introduction to the computer program *BERKELEY MADONNA*. *BERKELEY MADONNA* is designed to facilitate the process of dynamic modeling by providing various tools and algorithms for numerical solutions and graphical aids. A fully functional free download version (with only a few limitations regarding saving your models and printing) can be obtained from www.berkeleymadonna.com. The directions we give in this laboratory exercise refer to the commercial version.

We shall explore the capabilities of *BERKELEY MADONNA* on single-species population dynamic models. Because such models are relatively simple and they have been introduced and discussed in Chapter 1, we shall focus here primarily on the *BERKELEY MADONNA* syntax. Similar syntax is later used in more sophisticated laboratory projects.

In this lab, you will be asked to use *BERKELEY MADONNA* in order to examine some basis single-species population growth models. In doing so, you will learn how to perform the following basic operations in *BERKELEY MADONNA*:

- Enter and solve a discrete model.

- Enter and solve a continuous model.

- Perform batch runs for a set of values for a single model parameter.

- Vary the model parameters by defining and using sliders.

- Import and plot data sets.

- Perform curve fitting.

- Format the model output.

BIOLOGICAL BACKGROUND

The ability to reproduce is the hallmark of living organisms. The growth of a single-species population may be impacted by a variety of factors. These include the availability of adequate resources, climate conditions, predators and many others. In this laboratory exercise, we consider mathematical models for a single-species population that is developing under close to ideal conditions: We assume that unlimited resources are available, the absence of predators, and that death may only occur because of aging. Although unrealistic in general, these assumptions are often satisfied during the early phases of population growth in the absence of predation.

MATHEMATICAL BACKGROUND

In dynamical models, the quantities of interest change with time. Mathematically, we can describe such quantities by using function notation $p = p(t)$ to stress that the quantity p changes with time. Thus, $p(t)$ denotes the quantity at time t. Two different types of models will be introduced here and used throughout—discrete and continuous models.

Discrete models imply that change can only occur at specified time instances that are equally spaced. The length of the interval between two such consecutive "moments of change" is assumed to represent one unit of time. Thus, if we use a discrete model for the quantity $p = p(t)$ that changes in time, we will often use $p(0) = p_0$ to refer to the initial quantity (the quantity present when the process begins), and $p(1) = p_1$ to denote the quantity after the first change but before the second change. In general, $p(n) = p_n$ is used to denote the quantity p after the n-th change has taken place, but before the $n + 1$-st change has yet occurred. For example, if a discrete model is used to describe the yeast population growth data presented in Table 1-1 (Carlson [1913]; Pearl [1927]), $p(n) = p_n$ will denote the yeast biomass at the end of the n-th hour (i.e., $p_0 = 9.6$, $p_1 = 18.3$, and so on).

In Chapter 1, we built the following discrete population model

$$p_{n+1} - p_n = kp_n, \text{ or, equivalently, } p_{n+1} = (1 + k)p_n, \qquad (1\text{-}1)$$

which describes the change in population size at the end of each unit time interval as a (constant!) multiple of the size of the population in the beginning of the unit time interval. Model (1-1) is an example of a *difference equation*. In mathematical terms, a formula like this one that refers to previous terms to define the new term is called *recursive*. The value used to initiate the process is called *initial condition*. For this example, the initial condition is $p_0 = 9.6$.

Continuous models imply that an instant in time is equally as likely as any other to be a moment of change. Continuous models may often be presented in terms of conditions pertaining to the instantaneous rate of change exhibited by the dynamic quantity. Thus, continuous models are often given in terms of equations that involve the unknown function $p(t)$ and its derivative(s). Such equations are called *differential equations*. In Chapter 1, we examined the following continuous model of population growth:

Time (hours)	Observed Biomass
0	9.6
1	18.3
2	29
3	47.2
4	71.1
5	119.1
6	174.6
7	257.3

TABLE 1-1.
Yeast growth data, in thousands of cells per ml.

$$\frac{dP}{dt} = rP, r > 0. \qquad (1\text{-}2)$$

As in the discrete case, an initial condition $P(0)$ is necessary for finding the solution.

We showed in Chapter 1 that it is relatively easy to solve Eqs. (1-1) and (1-2) to obtain the exact (analytic) solution. However, as the sophistication of the models increases, the mathematical problem of solving the equations becomes increasingly more challenging. When it is difficult (or sometimes impossible!) to obtain the actual analytic solution, *numerical solutions* are used instead. A numerical solution does not give us an analytic expression for the unknown function but, instead, provides a table of values for this funtion. For example, a numerical solution of the problem , $\frac{dP(t)}{dt} = rP(t)$, $P(0) = 5.3$, for $r = 0.297$ is represented by Table 1-2.

The left column contains a grid of values for t, and the right column contains the values of the function $P(t)$ at these points. The increment used on the time variable t is often denoted by DT and is 0.5 in this example. Changing the value of DT allows for the creation of a specific grid of points at which the value of the function $P(t)$ will be calculated. For DT $= 1$, the time values for which $P(t)$ will be calculated will be $t = 0,1,2,3$, and so forth. For DT $= 0.2$, the time values will be $t = 0, 0.2, 0.4, 0.6$, and so forth.

There are various mathematical algorithms for solving and numerous software packages devoted to computing numerical solutions of differential equations. The one that we will be using throughout the book is called *BERKELEY MADONNA*. In this chapter, we refer to the specific syntax of *BERKELEY MADONNA*, although any other software package with similar capabilities can be employed instead.

The initial and final values of the time interval over which we would like to know the values of the solution should be specified. In *BERKELEY MADONNA*, they are called STARTTIME and STOPTIME. In the example above, we had the values of $P(t)$ calculated over the interval [0,6]. This will correspond to STARTTIME $= 0$ and STOPTIME $= 6$.

We now give a basic introduction that will allow you to enter mathematical models in *BERKELEY MADONNA* and obtain their numerical solutions. We shall use the models given as Eqs. (1-1) and (1-2) above as examples.

THE LANGUAGE OF *BERKELEY MADONNA*

The minimum information that one needs to provide in order to enter a model in *BERKELEY MADONNA* involves the following three items:

t (DT $= 0.5$)	$P(t)$
0.0	5.300
0.5	6.148
1.0	7.133
1.5	8.275
2.0	9.599
2.5	11.136
3.0	12.919
3.5	14.987
4.0	17.387
4.5	20.170
5.0	23.399
5.5	27.145
6.0	31.491

TABLE 1-2.
A numerical solution of the problem
$\frac{dP(t)}{dt} = rP(t)$, $P(0) = 5.3$, for $r = 0.297$.

1. An initial condition for the model;

2. The model itself in the form of a difference or a differential equation; and

3. Values for all parameters.

Although there is a slight difference in the syntax for entering difference and differential equations into *BERKELEY MADONNA*, the syntax for items 1 and 3 is the same regardless of the model type. It is important to stress here that *BERKELEY MADONNA*'s language is *not case-sensitive* (i.e., the program does not distinguish between *P* and *p*, *MODEL* and *model*, or *DiScReTe* and *discrete*).

The following tables exemplify the correspondence between the model description in mathematical terms and the corresponding *BERKELEY MADONNA* commands to be typed in the command window. More specifically, if we want to solve the discrete model from Eq. (1-1) with an initial condition of 9.6 and a value for the parameter of $k = 0.6$, we shall need to type in the commands that appear in the right column of Table 1-3.

If we want to solve the continuous model from Eq. (1-2) with an initial condition of 9.6 and a value for the parameter of, say, $r = 0.4$, we shall need to type in the commands that appear in the right column of Table 1-4.

You are now ready to explore *BERKELEY MADONNA* on your own!

Discrete Model	
Mathematical Description	*BERKELEY MADONNA* Commands
$p_0 = 9.6$ $p_{n+1} = (1 + k)p_n$ $k = 0.6$	`init` $p = 9.6$ `next` $p = (1 + k)*p$ $k = 0.6$

TABLE 1-3.
Mathematical description of the discrete model from Eq. (1-1) paired with the corresponding *BERKELEY MADONNA* syntax.

Continuous Model	
Mathematical Description	*BERKELEY MADONNA* Commands
$P(0) = 9.6$ $\dfrac{dP}{dt} = rP$ $r = 0.4$	`init` $p = 9.6$ `d/dt`$(p) = r*p$ $r = 0.4$

TABLE 1-4.
Mathematical description of the continuous model from Eq. (1-2) paired with the corresponding *BERKELEY MADONNA* syntax.

PART I: A DISCRETE MODEL

Start *BERKELEY MADONNA*. The Equations window appears with the following commands already in place (see Figure 1-1):

METHOD RK4

STARTTIME = 0

STOPTIME=10

DT = 0.02

All lines, except for the first one, should already be familiar. They specify that the mesh of time values for which the function values will be calculated begins at $t = 0$, ends with $t = 10$, and contain all points in between with increments of DT = 0.02. The first line specifies the numerical method that will be used by the program for computing the numerical solution for models described by differential equations (see Part II of this project). You can safely ignore this for now and accept the default algorithm. For the readers familiar with the theory of numerical methods for solving ordinary differential equations, we would add that *BERKELEY MADONNA* allows you to choose from a set of built-in algorithms, including Euler's method and two types of Runge–Kutta methods. More details on this and other specifics related to the software can be found in *BERKELEY MADONNA*'s brief documentation accessible under the Help menu.

The default values of STARTTIME, STOPTIME, and DT should be changed to the problem-specific values needed. For example, if we wish

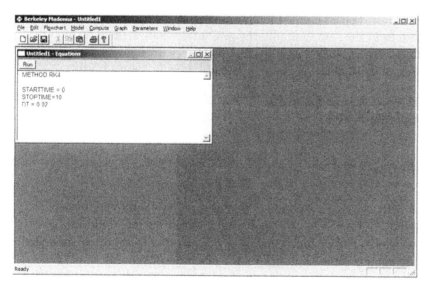

FIGURE 1-1.
The opening screen of *BERKELEY MADONNA*.

to obtain the numerical solution of the discrete model from Eq. (1-1) that corresponds to the yeast population data presented in Table 1-1, we can use STARTTIME = 0, STOPTIME = 7, and DT = 1.

Exercise 1-1

Entering a Model. Enter the discrete model from Table 1-3 into *BERKELEY MADONNA* by typing it, exactly as appears in Table 1-3, into the *Equations* window. Use the values STARTTIME = 0, STOPTIME = 7, and DT = 1 [see Figure 1-2(A)]. Run the model by clicking on the Run button. The graphical representation of the solution will appear in a new window, the *Graphics* window, as in Figure 1-2(B). To see the solution as a table of values, click on the Table button that is found across from the Run button in the graphics window—its icon is made of two squares offset from one another [Figure 1-2(C)]. The Table button is a toggle between graphical output and tabular output. Press the Table button again to go back to graphical output.

Exercise 1-2

Importing a Data Set. BERKELEY MADONNA can import data sets that have been saved in *plain text* format. You may enter the data using, for example, *Notepad* (separating the columns by a blank spaces or tabs) or type the data into an *MS Excel* spreadsheet and choose the Save As . . . Text (Tab delimited) option to save.

Enter the data from Table 1-1, and save as *YeastCulture1.txt*. To import this file into *BERKELEY MADONNA*, select File → Import Dataset from the main menu. Navigate to the file *YeastCulture1.txt*, and open it. Click OK in the Import Dataset dialog box [see Figure 1-3(A)]. The data should now appear on the plot. Depressing the Data Points button located across from the Run button in the graphical window (with an icon that features a solid black dot) will allow you to see the point on the graph that corresponds to the data points in Table 1-1, as in Figure 1-3(B).

How good do you think the fit between the predicted value and the actual values is?

Exercise 1-3

Changing Values in the Model. The model solution and goodness of fit depend on the values of the parameters used for the numerical calculations. For example, the value $k = 0.6$ was based on an estimate. Changing this value can improve or worsen the fit between actual and predicted values. If you wish to perform a run with a different value for a specified parameter, follow the procedure outlined below.

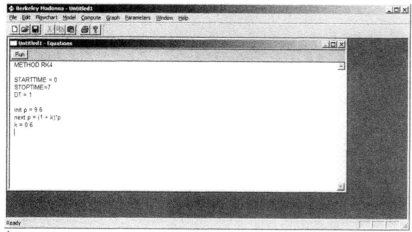

A

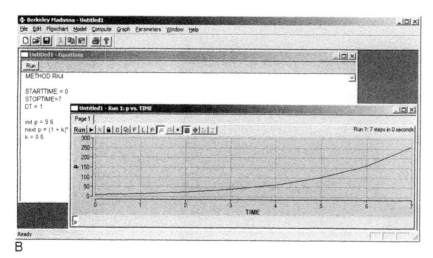

B

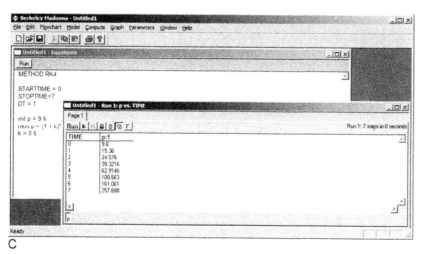

C

FIGURE 1-2.
Using *BERKELEY MADONNA*. Panel A: Entering the model; panel B: Result of the Run command;
panel C: Showing the table of values from the run.

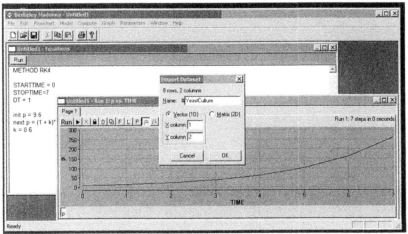

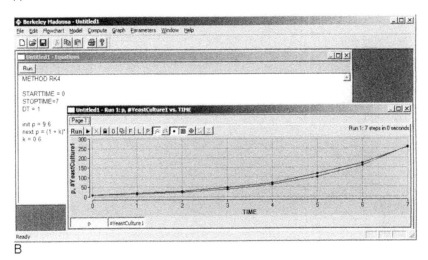

A

B

FIGURE 1-3.
Importing a dataset into *BERKELEY MADONNA*. Panel A: Using Import Dataset to import a text file; panel B: Viewing the dataset points on the graph.

1. From the *BERKELEY MADONNA* main toolbar, choose Parameters → Parameter Window.

2. The Parameters window appears (see Figure 1-4).

3. You can now change the value of any parameter by highlighting it in the window and typing in a new value. You can change as many values as you like

4. Click the Run button in the parameter window to run the model with the new set of parameter values.

Run the model with the following values for k: $k = 0.55$; $k = 0.50$; $k = 0.45$. Comment on whether the fit between predicted and actual values improves. If at any time you want to change the value of

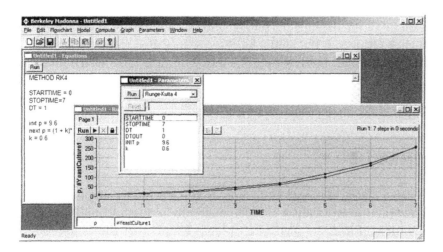

FIGURE 1-4.
Using the Parameters window to alter the parameter values.

a parameter back to its original value given in the Equations window, highlight it in the Parameters window and click the Reset button.

Note. If the *Parameters* window hides behind some of the other windows and you want to use it, it can be brought back by choosing Window from the main toolbar and then choosing Parameters from the drop-down menu.

Exercise 1-4

Performing Batch Runs. Rerunning the model with different values for k is a tedious job. Moreover, it may be difficult to compare runs corresponding to different values of k when the outputs appear on different graphs. When using batch runs, we can specify increments and ranges for the parameter values, then perform a run that will display all of the solutions on a single graph. Here is the procedure:

1. Select Parameters → Batch Runs from the main toolbar.

2. The Batch Runs window will appear.

3. From the pull-down menu Parameters, select k.

4. Enter 5 in the Textbox labeled "# of Runs."

5. Give Initial Value of 0.4 and Final Value of 0.8.

Because you specified 5 as the number of runs, *BERKELEY MADONNA* will now use the value k = 0.4 as an initial value, the value k = 0.8 as the final value, and will divide the interval [0.4,0.8] into four equal subintervals to determine the remaining three values for k that should be used. Those values appear in the Values window.

Accept all of the defaults for the remaining components of the window, and click OK to run the model for the specified five values of k. Click the Legend button in the graphics window to determine which lines correspond to which values of k (it can be found across from the Run button and is labeled L). Describe what effect the increase of k has on the solution of the model. Give a biological justification for your observation.

Exercise 1-5

Defining a Slider. You now know how to run the model simultaneously for several different values of a parameter. In addition to supporting batch runs, *BERKELEY MADONNA* has a feature that allows for real-time viewing of the changes to the solution caused by changes of the model parameters. All you need to do is define a slider.

1. Select Parameters → Define Sliders from the main toolbar.

2. The Sliders window will appear.

3. From the parameter list on the left select k by highlighting.

4. Click Add to add k to the list of parameters for which you wish to define a slider.

5. The parameter k will now appear in the list on the right, under Sliders: [see Figure 1-5(A)].

6. Accept the default values for the range of values of k and the magnitude of the increment, or introduce values of your own choice. Click OK.

7. The slider will now appear. Click on the arrows to see how changing k affects the model.

8. Checking the 10X checkbox that appears on the slider refines the increment by a factor of 10.

Use the slider to determine the value of k that provides the best visual fit between the model predictions and the data in *YeastCulture1.txt*. State this value.

Exercise 1-6

Performing a Curve Fit. Indisputably, inspecting the graph to determine the best visual fit between predicted and actual values is not the best strategy. What appears to be the best fit to you may not look as good a fit to your lab partner. The only objective way to determine who is right and is to give a precise definition of term *best fit*. Mathematically, this can be done by defining a function that measures the combined

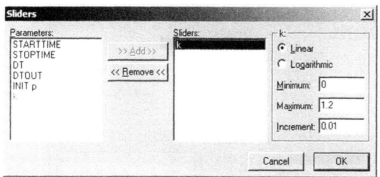

A

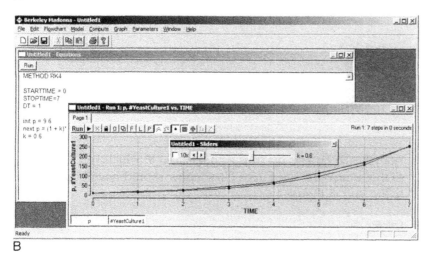

B

FIGURE 1-5.
Using the slider function to adjust parameters. Panel A: Defining a slider with the Sliders window; panel B: The slider is moved to the left or the right to alter the parameter value and change the model.

deviation between actual and model-predicted values. The best fit is then achieved for the parameter values that minimize this deviation. We present the relevant mathematical theory in Chapter 8 of the textbook. In this introduction, we shall simply demonstrate how to use the appropriate _BERKELEY MADONNA_ feature to calculate the best fit for the model parameters. These are the steps you need to follow:

1. Select Parameters → Curve Fit from the main toolbar.

2. The Curve Fit window will appear.

3. From the Available list, select the parameter to be estimated by highlighting it (in this case, select k).

4. Click Add, and observe that k appears in the Parameters list.

5. Select P for Fit Variable (this is the model variable whose predicted values are compared with the actual data), and _YeastCulture1_

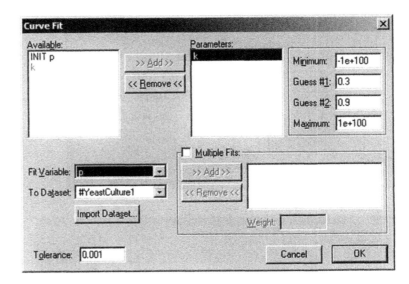

FIGURE 1-6.
The Curve Fit window.

from the To Dataset list (this is the dataset to which you would like to fit the model). The Curve Fit window should now look similar to the that in Figure 1-6.

6. Click OK.

Report the value of k that provides the best fit to the data. You can obtain this value either from the Sliders window or from the Parameters window.

Exercise 1-7

Exporting the Program Output. It is often necessary to export the results from your *BERKELEY MADONNA* calculations for use by other applications. For example, it may be useful to read the table containing the model predictions and numerical solution into a spreadsheet application, or to export the graph of the solution as a (picture) file that will then be appropriate to insert into other documents.

To save the table with the model output:

1. Switch to Table View mode by depressing the Table button.

2. From the main toolbar, select File → Save Table As. A file dialog will appear.

3. Navigate the file dialog to the folder in which you want to save the file.

4. Specify the name under which you wish to save the file. Click Save.

To save the graph(s) of the model:

1. Switch to Graph Mode by releasing the Table button.

2. From the main toolbar, select File → Save Graph As. A file dialog will appear.

3. Navigate the file dialog to the folder in which you want to save the file.

4. Specify the name under which you wish to save the file. Click Save.

For the value of k determined in Exercise 6, save the Table and the Graph containing the predicted and actual values. Present the Table and Graph as part of your lab report.

PART II: CONTINUOUS MODELS

Exercise 1-8
....

Entering the Model. We will continue to work with the yeast data from Table 1-1, but, in this section, we will consider the continuous model expressed by Eq. (1-2) to describe it. Start *BERKELEY MADONNA*, and type the equations for the model exactly as they appear in the right column of Table 1-4. Use STARTTIME = 0, STOPTIME = 7, and DT = 1, as we did for the discrete model. Click Run. The Graphics window will appear as in Figure 1-7.

Using the techniques learned from Part I, perform the following tasks:

(a) Import the file *YeastCulture1.txt* to obtain plot of the data.

(b) Click on the Data Points button (the button labeled with a solid dot in the Graphics window) to see the actual data points and model-calculated values. How good is the fit?

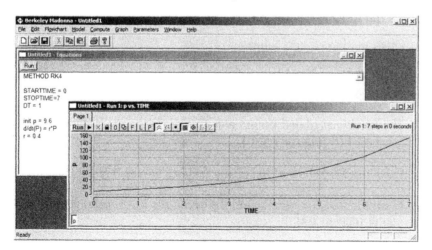

FIGURE 1-7.
Entering and running a continuous model.

(c) Define a slider for r, and use it to determine the best visual agreement between the actual data points and the values predicted by the model. State that value.

(d) Use the Curve Fit option of *BERKELEY MADONNA* to determine the value of the parameter r that provides the best fit between predicted and actual values for the given initial condition. State that value.

Exercise 1-9

Using a mathematical model to estimate population size at times unavailable from experimental data. The experimental data from Table 1-1 provide information for observed biomass at the end of each hour from the beginning of the data collection, but no intermediate values are measured. Once we have determined the best fit between predicted and measured values as in Exercise 8(d), we can use the model to generate approximate values for the population size at times unavailable from the experiment. Assume, for example, that we wish to assess the yeast biomass at time $t = 3.5$ hours [i.e., we want to find an estimate for $P(3.5)$].

To obtain the value $P(3.5)$, we need to make sure that the value $t = 3.5$ is among the set of tabulated values for t. Thus, we need to choose DT appropriately. For example, DT = 0.5 or DT = 0.25 will do the job, because when we start form $t = 0$ and use either value of DT, the point $t = 3.5$ will be one of the tabulated values. On the other hand, DT = 0.2 will not be appropriate, as when we start from $t = 0$ with DT = 0.2, we will skip the value $t = 3.5$.

As for the discrete model, changing the value of DT can be done by using the Parameters window (see Figure 1-8). The following list of steps outlines the *BERKELEY MADONNA* commands:

1. Open the Parameters window.

2. Highlight DT in the Parameters window, change its value to 0.5 [see Figure 1-8(A)], and run the model again.

3. The newly calculated points will appear on the graph of the function $P(t)$ [Figure 1-8(B)].

4. Press the Table button to see the table of values [Figure 1-8(C)].

5. From the table of values, find the value for $P(3.5)$ predicted by the model [Figure 1-8].

Use the same continuous model to find approximations for the values $P(5.3)$, $P(2.75)$, and $P(6.8)$.

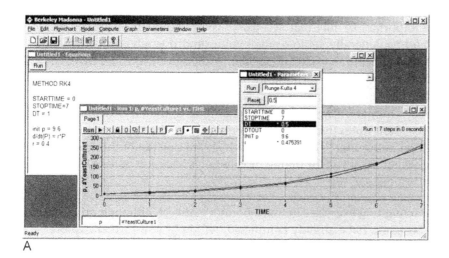

A

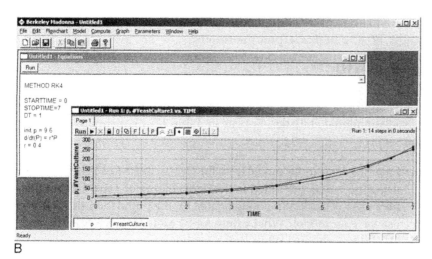

B

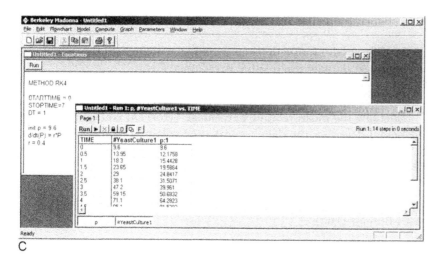

C

FIGURE 1-8.
Obtaining information from the model. Panel A: Decreasing the value of the parameter DT; panel B: Data points corresponding to the new DT are now visible on the graph; panel C: The new points are shown on the table of values.

It is important to note that because of computational considerations, DT should be generally set to a small value, and it is advisable to use the default value of DT = 0.02 or smaller. Although *BERKELEY MADONNA* is fully capable of performing the calculations with any user-entered DT value, larger values may introduce significant inaccuracies in the model's numerical solution. The magnitude of the error depends to a large extent upon the particular model and on the specific numerical method being used, the details of which are outside of the scope of our discussion. As a rule of thumb, however, *we shall use values* DT = 0.02 *or smaller from now on.*

The remaining exercises in this laboratory project, although quite similar to those you have already completed, will help you further explore *BERKELEY MADONNA*. There are more decisions in these exercises that you will have to make on your own, such as choosing appropriate values for STARTTIME, STOPTIME, and DT. We also expect you to carefully consider how to best interpret the results.

Exercise 1-10

Census data from Statistics Sweden, the Swedish government authority for official national statistics, appear in Table 1-5 (Statistics Sweden 2003).

Year	Swedish Population (thousands)
1760	1,925
1780	2,118
1800	2,347
1820	2,585
1840	3,139
1860	3,860

TABLE 1-5.
Swedish population data.

(a) Enter the continuous model from Eq. (1-2) in *BERKELEY MADONNA*, using the syntax presented in the right column of Table 1-4, but this time use $r = 0.004$. What initial value for the population size will you use for your model? What values for STARTTIME, STOPTIME, and DT did you choose? Why? Run the model.

(b) Determine the value of the parameter r that provides the best fit between predicted and actual values. Report the value of r.

(c) List the model predictions, and compare them with the actual values. Calculate the relative percentage error
$$(\% \text{ error} = \frac{|observed - predicted|}{observed} 100)$$ at each time point and calculate the mean relative percentage error. Do you think that the model describes the data accurately? Explain your conclusion.

Exercise 1-11

Repeat Exercise 10 with the U.S. population data from Table 1-6 (U.S. Census Bureau Data).

Year	U.S. Population (millions)
1800	5.3
1810	7.2
1820	9.6
1830	12.9
1840	17.1
1850	23.2
1860	31.4

TABLE 1-6.
U.S. population data.

Exercise 1-12

Were the values of r that provided the best fit to the data from Exercises 10 and 11 the same or different? Give an explanation.

Exercise 1-13

Use both the discrete model from Eq. (1-1) and the continuous model from Eq. (1-2) to predict the population of:

1. The United States for the year 2000; and

2. Sweden for the year 2000.

According to U.S. Census figures, the U.S. population in 2000 was 281.4 million. The population of Sweden for 2000 was 8.883 million (Statistics Sweden 2003). Compare the models' predictions with the actual figures. Calculate the percentage error, and comment on the accuracy of the model prediction.

Exercise 1-14

(a) From a mathematical point of view, why are the models (1-1) and (1-2) not completely realistic? What mathematical property of the solutions is causing this?

(b) From a biological point of view, why are the models (1-1) and (1-2) not completely realistic? Which are the biological assumptions we made that are causing this?

REFERENCES

Carlson, T. (1913). Über Geschwindigkeit und Grösse der Hefevermehrung in Würze. *Biochemische Zeitschrift, 57*, 313–334.

Pearl, R. (1927). The growth of populations. *Quarterly Review of Biology, 2*, 532–548.

Statistics Sweden (2006, February 14). *Population and population changes 1749–2005.* Retrieved September 25, 2006, from http://www.scb.se/templates/tableOrChart____26047.asp

U.S. Census Bureau (1993, August 27). *Selected historical decennial census population and housing counts. Population, housing units, area measurements, and density: 1790 to 1990.* Retrieved September 21, 2006, from http://www.census.gov/population/censusdata/table-2.pdf

U.S. Census Bureau (2001, April). *Population change and distribution, 1990 to 2000. Census 2000 brief.* Retrieved September 25, 2006, from http://www.census.gov/prod/2001pubs/c2kbr01-2.pdf

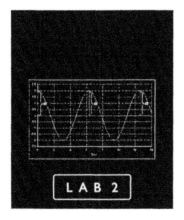

Logistic Models of Single-Species Population Dynamics, Equilibrium States, and Long-Term Behavior

In Lab 1, we saw examples of populations that seem to grow at a constant per capita rate for some period of time. If, however, we assume that this growth pattern continues indefinitely, we see results that are obviously impossible.

We begin this lab with analyzing data from the growth of a yeast culture. We use the data to guide us in choosing a more realistic model. Yeast would seem to be a good choice because it goes through many generations rapidly and is easy to isolate in a closed system (i.e., a petri dish).

In this laboratory exercise, we refine the previously explored discrete models and introduce a class of models called logistic models. A defining characteristic of these models is the assumption that populations do not grow at a uniform per capita rate as assumed by Malthus in his 1798 work *An Essay on the Principle of Population* (Malthus [1798]), but that this rate changes with the size of the population. Verhulst, who published his results in 1845, explored the idea in mathematical terms, and arrived at a solution that involved a logarithm. Verhulst used the term *logistic* for the solution, most likely intending to stress that it was given by a "log-like" curve. We examine both discrete and continuous logistic models.

We also examine the question of equilibrium states of a population and investigate long-term behavior of the model solutions and convergence to equilibrium. Equilibrium states are important because many systems in nature stabilize, in the long run, around an equilibrium state. This means that if we wait long enough, the motion for such systems eventually stops, after a state is reached that is perceived by the system to be optimal in some sense. This, of course, is by no means true for all systems that evolve with time.

In this lab, you will:

- Modify the unlimited population growth models to make them better describe experimental data.

- Understand the meaning of equilibrium states.

- Learn how to find the equilibrium states for a dynamic model.

- Derive information about the model trajectory without solving the model.

- Determine the dependence of model trajectories on initial conditions.

BIOLOGICAL BACKGROUND

Why don't populations of organisms grow in an unlimited fashion in the real world? If a mathematician were to ask this question of a biologist, the answer would probably include the words "carrying capacity." Simply put, carrying capacity is the number of organisms that the environment can support. The carrying capacity represents the theoretical upper limit on the size of a particular population that a particular environment can sustain. When the population size has reached the carrying capacity, there should be no more net growth in the size of the population.

Many factors contribute to the observed limitation of population growth. The availability of resources, such as food and water, shelter, and nesting sites, will limit growth, as will pollution or other degradation of the environment. Ecologists group these factors into two groups—those that depend on population density (density-dependent factors) and those that do not (density-independent factors). The food supply would represent a density-dependent factor—as the population grows, food becomes limited, and some members of the population may starve to death. The birth rate may also fall as the effects of malnutrition are manifested in decreased fertility. A density-independent factor might be an unusually severe winter storm, because the weather is not a consequence of the population density.

When a population is at its carrying capacity, should we expect it to remain there indefinitely? In the wild, a healthy ecosystem might be expected to maintain a given number of organisms of each species. In such a case, the birth rate would approximately balance the death rate, resulting in a relatively stable population. If the population should rise above the carrying capacity (perhaps because of the immigration of a number of individuals into the particular area under study), we would expect the death rate to rise and the population to fall until it reaches the carrying capacity again.

In the laboratory, however, we often observe a different result. In a typical bacterial population, for example, after the carrying capacity has been reached, the organisms will eventually begin to die, because the resources are not being continuously replenished nor is the waste being removed. Interestingly, death occurs exponentially, and this population decline phase is the mirror image of the exponential growth phase of the culture.

MATHEMATICAL BACKGROUND

The calculus concepts that will be used in this laboratory exercise are limit values, asymptotic behavior, and use of derivatives for constructing sketches of functions' graphs.

Recall that if $P(t)$ is a function with function values that get closer and closer to a specific fixed value K as the independent variable t increases without a bound, we say that the function $P(t)$ has K as a *limit* and write $K = \lim P(t)$, as $t \to \infty$. We also say that the value K is a horizontal *asymptote* for the function $P(t)$.

The functions of interest for this laboratory project will be solutions of difference or differential equations. In this context, the graphs of the solutions are often referred to as *time trajectories,* or simply *trajectories*. The following dependence exists between the growth/decrease of a trajectory and the sign of the function's derivative: If the derivative is positive over a certain time interval, the function increases over this interval; a negative value of the derivative over an interval indicates that the trajectory is decreasing there. When the derivative is zero, the trajectory remains flat (i.e., there is no change).

NECESSARY SOFTWARE

The software necessary for this lab is *BERKELEY MADONNA*. Lab 1 should be completed as a prerequisite.

Exercise 2-1

Table 2-1 contains yeast culture growth data obtained during an 18-hour period. Biomass was measured at the end of each hour (Carlson [1913]; Pearl [1927]).

Enter the data and save it in a file *YeastCulture2.txt* (every row in this file should contain only two numbers: the time and corresponding biomass value).

   (a) Import the file *YeastCulture2.txt* to *BERKELEY MADONNA* to plot the data.

Time t	Biomass $P(t)$	Time t	Biomass $P(t)$	Time t	Biomass $P(t)$
0	9.6	7	257.3	14	640.8
1	18.3	8	350.7	15	651.1
2	29	9	441	16	655.9
3	47.2	10	513.3	17	659.6
4	71.1	11	559.7	18	661.8
5	119.1	12	594.8	—	—
6	174.6	13	629.4	—	—

TABLE 2-1.
Yeast culture growth data obtained during an 18-hour period.
(Data taken from Carlson [1913] and Pearl [1927].)

(b) Describe the pattern of the data. When does it increase most rapidly?

(c) Do you think biomass would ever grow to be as large as 700?

Exercise 2-2

Do the data of Table 2-1 suggest that there is a number K that the biomass approaches as time grows indefinitely? If so, give an estimate for this value.

LOGISTIC GROWTH MODELS

The model we would like to develop here is a modification of the unlimited growth model

$$\frac{dP}{dt} = rP, r > 0. \tag{2-1}$$

Recall that for this model, $P(t)$ is the population size at time t, and the parameter r, representing the net per capita rate of growth for the population, is constant. Because it provided a fair fit with experimental data during the initial and most rapid phase of growth, we would like to have this dependence preserved (or preserved to a large extent) for the specified range. Because it failed to provide an adequate fit with the experimental data during the later phases of growth, we would like to modify it and improve the performance of the model there. We approach the problem by following the idea of Verhulst that the net per capita rate of growth will no longer be constant but will now depend on the size of the population. Thus, r is now a variable that depends on $P(t)$. Therefore, Eq. (2-1) is now better written as

$$\frac{dP}{dt} = r(P(t))P(t). \tag{2-2}$$

With this modification, the model should be able to describe a process that reaches a "level of saturation" that corresponds to the population's carrying capacity. The important question here is to determine the specific functional dependence $r(P(t))$. We shall describe some characteristics that our model should have, and then develop a set of equations that describe them. There is not a unique "correct solution" here, and considering alternative dependences for $r(P(t))$ is certainly worthwhile.

Exercise 2-3

Let K denote the maximum number of organisms the environment can sustain. Suppose that a number of organisms are introduced into the environment so that the total number of organisms within the environment at time $t = 0$ exceeds K. What should then happen to the population?

Exercise 2-4

Suppose that we want the rate of growth to fluctuate so that if the environment has fewer than K of the organisms, the population grows in size, and if the environment has more than K of the organisms, the population decreases in size. Describe what properties, if any, this will force dP/dt to have. Describe what properties, if any, this will force $r(P(t))$ to have.

Exercise 2-5

Suppose the maximum per capita rate of growth is a. Give an equation for $r(P(t))$ that has all of the following characteristics:

- $r(P(t))$ is largest when $P(t) = 0$.

- If $P(t) < K$, then $r(P(t)) > 0$.

- If $P(t) = K$, then $r(P(t)) = 0$.

- If $P(t) > K$, then $r(P(t)) < 0$.

Exercise 2-6

(a) What does the condition $r(P(t)) = 0$ when $P(t) = K$ say about the growth/decline of the population when $P(t) = K$?

(b) What does the condition $r(P(t)) > 0$ when $P(t) < K$ say about the growth/decline of the population when $P(t) < K$?

(c) What does the condition $r(P(t)) < 0$ when $P(t) > K$ say about the growth/decline of the population when $P(t) > K$?

Exercise 2-7

Does the model below satisfy all conditions from Exercise 2-5?

$$\frac{dP}{dt} = \frac{a}{K}(K - P)P \qquad (2\text{-}3)$$

Explain why or why not. The model defined in Eq. (2-3) is a *logistic population growth model*.

Exercise 2-8

Enter the logistic model from Eq. (2-3) into *BERKELEY MADONNA*.

(a) Choose a value for K based on the data plot (see also Exercise 2-2).

(b) Choose a value for a to start with. Run the model. Define a slider for the parameter a to determine a value for this parameter that fits the data best. What is this value?

Exercise 2-9

Remove the data set *YeastCulture2* from the plot. We now investigate in what way the parameter a and the initial condition $P(0)$ affect the logistic model from Eq. (2-3).

(a) Using a slider for the parameter a, describe the effect that increasing the value of a has on the way the population evolves.

(b) If $K = 1000$, $a = 0.1$, and $P(0) = 1500$, what behavior would you expect for the model? Why?

(c) Run the model with $K = 1000$, $a = 0.1$, and $P(0) = 1500$. Describe what happens when the initial size of the population exceeds the carrying capacity of the environment.

Exercise 2-10

Compare the performances of the logistic model from Eq. (2-3) and the model from Eq. (2-1). Comment on the advantages and disadvantages of the logistic model. Give any suggestions for modifying the logistic model that may improve it further.

Exercise 2-11

Following the same steps as in the continuous case, modify the discrete model $P_{n+1} - P_n = k P_n$ (replacing the constant k with an appropriate function $k(P(t))$ to obtain a discrete logistic model of population growth. Find the values of the model parameters providing the best fit for the data in Table 2-1. Include with your laboratory report the model modifications and the model output for the parameter values that provided the best fit. We shall see later that although this discrete model appears similar to the continuous model from Eq. (2-3), it exhibits a fundamentally different long-term behavior.

EQUILIBRIUM STATES, STABILITY, AND LONG-TERM BEHAVIOR

When examining the behavior of a dynamic quantity $P(t)$, the following questions are of particular interest:

- What happens with this quantity in the long run, and will it reach a limit level K and stabilize around it?

- Will the answer to the previous question depend in any way on the initial condition $P_0 = P(0)$ and/or the specific values of the model parameters. If so, to what extent?

- Are there values P_0 for the population size $P(t)$ such that if $P(0) = P_0$, the population size remains equal to P_0 at all times? That is, can we find values P_0 for the population size $P(t)$ such that $P(t) = P_0$ for all t?

Values that satisfy the condition $P(t) = P_0$ for all t are called *equilibrium states*. Let's consider the question of determining the equilibrium states first.

Exercise 2-12
....

By the definition we gave, the equilibrium states for the system are the ones for which the population size remains unchanged at all times. Give the analytic condition that determines the equilibrium states, but do not yet determine those equilibrium states. Consider the cases of discrete and continuous models separately.

Exercise 2-13
....

What are the equilibrium states for the model $\dfrac{dP}{dt} = rP$, if $r \neq 0$?

Exercise 2-14
....

Find the equilibrium states for the following population growth models:

(a) $\dfrac{dP}{dt} = a(1 - \dfrac{P}{K})P$, where $K > 0$ and $a > 0$ are constants.

(b) $P_{n+1} = P_n + \dfrac{a}{K}(K - P_n)P_n$, where $K > 0$ and $a > 0$ are constants.

(c) What happens if we consider $a = 0$ in the models from (a) and (b)?

Our next exercise demonstrates that change in the long-term behavior may develop for certain values of the parameters. For example, a trajectory may converge to equilibrium for certain parameter values and may oscillate for other ones. Specifically, we consider a version of the discrete logistic growth model from Exercise 14(b) above. In this model, denote $x_n = \dfrac{P_n}{K}$, so that x_n is the current population expressed as a fraction of the maximum population the environment can sustain.

Exercise 2-15
....

Show that with the above notation, the Verhulst model from Exercise 14(b) can be written in the form

$$x_{n+1} = x_n + a(1 - x_n)x_n \qquad (2\text{-}4)$$

and that the "carrying capacity" of the model from Eq. (2-4) is equal to 1.

Our next exercise shows that as the values of the parameter a for this model change, the long-term behavior of the trajectory changes from one range of a to the next. For large values of a in this model, the long-term behavior could be described as *chaotic*. See Chapter 1 for more details.

This brings us again to the question of how continuous and discrete models could differ. Although the discrete and continuous models have similar mathematical form, their behaviors can be radically different. Thus, it should not be assumed that when a continuous model describes a process accurately, its discrete analogue would describe the process equally well. Instead, the modeling process and this choice should be based strictly upon the specific biological problem at hand.

Exercise 2-16

Enter the model from Eq. (2-4) in *BERKELEY MADONNA*. Run the model with an initial condition $x_0 = 0.5$, and define a slider on a for the range from $a = 0$ to $a = 3$, with increment 0.1. Using the slider, examine the plots as well as the table of values as a increases. To decrease the increment for the slider when necessary, check the 10x box.

Hint: When the values of a increase sufficiently, look for periodic behavior and the length of the repeating sequence of values.

Answer the following questions:

- What is the long-term behavior of this model for $0 < a \leq 1.0$?

- What is the long-term behavior of this model for $1.0 < a < 2$?

- What happens for $a = 2$? Why?

- What is the long-term behavior of this model for $2.01 \leq a \leq 2.44$?

- What is the long-term behavior of this model for $2.45 \leq a \leq 2.54$?

- What is the long-term behavior of this model for $2.55 \leq a \leq 2.56$?

- What is the long-term behavior of this model for $2.56 < a < 3$?

Repeat the exercise with different values for the initial condition x_0. Do you think that the long-term behavior of the model depends upon the choice of the initial condition x_0?

Exercise 2-17

We close with the comment that it is by no means the case that all biological systems reach equilibrium. Can you think of some biological systems that approach equilibrium states with time and others that do not exhibit such behavior? Give examples.

REFERENCES

Carlson, T. (1913). Über Geschwindigkeit und Grösse der Hefevermehrung in Würze. *Biochemische Zeitschrift, 57,* 313–334.

Pearl, R. (1927). The growth of populations. *Quarterly Review of Biology, 2,* 532–548.

Malthus, T. R. (1798). *An essay on the principle of population.* London: Printed for J. Johnson, in St. Paul's Church-Yard. Retrieved September 25, 2006, from http://www.ac.wwu.edu/~stephan/malthus/malthus.0.html

FURTHER READING

Verhulst, P. F. (1845). Recherches mathématiques sor la loi d'accroissement de la population. *Nouveaux Mémoires de l'Académie Royale des Sciences et Belles-Lettres de Bruxelles, 18,* 1–38.

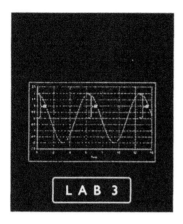

Physiological Mechanisms of Drug Elimination from the Bloodstream and Optimal Drug Intake Regimens

Why do you need to take two acetaminophen tablets every 4 to 6 hours when you have a headache? Why is there a warning on the label that cautions you not to take any more than four doses in a given 24-hour period? And why does your head start to ache again after 4 hours when the above warning suggests that you really ought to wait until 6 hours have elapsed before you take the next dose? These questions serve to introduce our next topic, that of modeling the concentration of physiologically active substances in the bloodstream.

For both prescription and over-the-counter drugs, the recommended dose regimen is designed with the following important aims in mind: Ensuring that the dosage is high enough to provide the desired effect, and ensuring that the dosage is not so high that the drug is toxic. Two critical drug concentrations are thus essential—the minimum effective concentration (MEC) and the minimum toxic concentration (MTC). Between these two limits lies the therapeutic window—the range of concentrations over which the drug is both effective and safe. The MEC and MTC are drug-specific and determine the framework for designing optimal drug intake schedules.

The optimal schedule will naturally depend upon the way the drug concentration in the blood changes with time, and determining the exact dynamics is not trivial. At any time, the concentration is the result of two competing simultaneous processes: Concentration increase caused by drug intake, and concentration decrease caused by the physiologically provoked elimination of the drug from the bloodstream. A mathematical model could facilitate significantly our understanding of the evolution of drug concentration dynamics and help the design of optimal drug delivery schedules.

For practical reasons, it is easier to follow a schedule in which equal doses are given at regular time intervals. Although this consideration may certainly be ignored in emergency treatments, for the vast majority of drugs it has become the norm for designing drug intake schedules. Thus, our problem for this laboratory exercise could be summarized as follows:

Knowing the MEC and the MTC for a specific drug and how fast it is eliminated from the bloodstream, determine the amount of each dose and the length of time between doses to ensure a concentration that reaches the therapeutic window rapidly and remains there for the duration of the treatment.

In this lab, you will:

- Construct a model that describes the drug concentration dynamics.

- Examine the cumulative effect of multiple doses.

- Examine the effect caused by changing the amount given with each dose and the time between doses.

- Given the MEC and MTC, determine an appropriate dose regimen.

BIOLOGICAL BACKGROUND

The factors that influence the concentration of a drug once it is in the bloodstream may be roughly divided into factors controlling entry of the drug into the bloodstream and factors controlling clearance of the drug from the bloodstream.

In pharmacology, factors that control the entry of the drug into the bloodstream are grouped under the general term *absorption*. Knowing by what route the drug will be introduced into the patient, either orally, intravenously, intramuscularly, transdermally (through the skin), or by inhalation, is crucial for determining absorption. It is also necessary to consider whether the drug undergoes any physical or chemical changes during the administration. For example, will it be administered as a solid that must be dissolved in the stomach or small intestine, or will it be introduced as a solution? The characteristics of the drug that control the manner in which it will be absorbed into the bloodstream should also be considered. A fat-soluble molecule can diffuse through the intestinal cell membranes, whereas a charged or polar molecule would have to rely on transport proteins in the cell membrane.

Once the drug has entered the bloodstream, we must consider how the drug is dispersed throughout the body. The pharmacologist's term for this issue is *distribution*. For example, does the drug bind to serum proteins, or does it circulate as free drug? Distribution also involves the movement of the drug from the bloodstream to the tissue, organ, or other area of the body where it will have an effect.

Although some of the drug may never get to the bloodstream to have an effect on your headache (e.g., a portion of the acetaminophen will remain in and leave the body through the digestive tract), the remainder will gradually be removed from the bloodstream by processes grouped under the term *elimination*. Elimination is most often caused by the action of the kidneys, but metabolism may also be involved.

The kidneys filter the blood and remove drugs and metabolic wastes. However, drugs may also be excreted by the liver in the bile. Bile, a substance necessary for the digestion of fats, is made by the liver, stored in the gallbladder, and delivered into the small intestine. Drugs may also be removed from the bloodstream by the lungs and then exhaled. (If this sounds unlikely, just think of the instruments commonly used for testing drivers suspected of driving under the influence of alcohol.)

Metabolism (the term for any chemical reaction the drug may undergo in the body) occurs through the action of protein catalysts called enzymes. Metabolism of drugs most often occurs in the liver, though there are some enzymes in the bloodstream as well. Liver enzymes involved in drug metabolism include epoxide hydratase and the cytochrome P450 family. Liver enzymes catalyze the chemical modification of the drugs, often oxidizing them and making them more amenable to excretion.

By this point, it should be clear that we would need to begin our model construction by making some simplifying assumptions.

The simplest model would be to assume an instantaneous entry of drug into the bloodstream, followed by gradual clearance from the bloodstream. An intravenous injection of the drug might well approximate the instantaneous entry. How might we model the clearance of the drug from the bloodstream? Generally, the rate of clearance of drugs from the bloodstream is proportional to the amount that is present in the bloodstream. Therefore, we are looking at an example of exponential decay—the reverse of our first population model.

If $C(t)$ is the concentration of a drug at time t, then the fact that the drug is eliminated from the bloodstream at a rate proportional to the amount present can be expressed as

$$\frac{dC(t)}{dt} = -rC(t), \qquad (3\text{-}1)$$

where $r > 0$.

The negative sign on the right-hand side of Eq. (3-1) indicates that the derivative $\frac{dC}{dt}$ of the concentration function $C(t)$ is negative, and thus the concentration of the drug in the blood is decreasing.

The constant r, called *elimination rate constant*, controls the rate at which the drug is removed from the blood. It is closely related to the *half-life* of the drug, defined as the time necessary to reduce the concentration of the drug in the blood by 50%. In mathematical terms, the half-life t^* is the time elapsed since the initial moment $t = 0$, for which $C(t^*) = 0.5C(0)$. Because the analytic solution of Eq. (3-1) is $C(t) = C(0)e^{-rt}$, we obtain $C(0)e^{-rt^*} = 0.5C(0)$, so $e^{-rt^*} = 0.5$. Thus $-rt^* = \ln(0.5)$, leading to the following connection between the elimination rate constant r and the half-life of the drug:

$$t^* = \frac{\ln(2)}{r}.$$

MATHEMATICAL BACKGROUND

The basic mathematical concepts in this laboratory exercise involve geometric series, graphs of exponential functions, step functions, and function translations and truncations.

Geometric Series. An expression of the form

$$a + ab + ab^2 + ab^3 + \ldots$$

is called a geometric series. If we denote the sum of the first n terms of such a series by S_n, that is:

$$S_n = a + ab + ab^2 + ab^3 + \ldots + ab^{n-1},$$

then

$$bS_n = ab + ab^2 + ab^3 + \ldots + ab^{n-1} + ab^n,$$

and

$$S_n - bS_n = (1-b)S_n = a - ab^n.$$

Thus, if $b \neq 1$, we obtain the following compact formula for the sum S_n of the first n terms of a geometric series:

$$S_n = \frac{a - ab^n}{1-b} = a\frac{1 - b^n}{1-b}. \tag{3-2}$$

When $|b| < 1$, we have:

$$\lim_{n \to \infty} S_n = \frac{a}{1-b}. \tag{3-3}$$

The geometric series that occur in this module will be of the form:

$$C + Ce^{-Tr} + C(e^{-Tr})^2 + C(e^{-Tr})^3 + \ldots.$$

Applying Eq. (3-2) with $a = C$ and $b = e^{-Tr}$, where C will be the dosage and T the time between doses, we obtain:

$$C + Ce^{-Tr} + C(e^{-Tr})^2 + C(e^{-Tr})^3 + \ldots + C(e^{-Tr})^{n-1} = C\frac{1 - (e^{-Tr})^n}{1 - e^{-Tr}}. \tag{3-4}$$

We now recall some facts that will enable us to graph the concentration of the drug.

Graphs of Functions, Translations, Step Functions, and Truncations. Recall that if the graph of a certain function $f(t)$ is known, the graph of the function $f(t - c)$, where $c > 0$ is a constant, can be obtained from the graph of $f(t)$ by horizontal translation of c units in the positive direction of the t-axis. For example, knowing the graph of $f(t) = 5e^{-2t}$, the graph of $g(t) = f(t - 1) = 5e^{-2(t-1)}$ in Figure 3-1 is obtained by shifting it one unit to the right.

The function $f(t) = 5e^{-2t}$ is defined for all values of t—positive and negative. For applied problems, where the variable t usually represents time, it is often appropriate to use the truncated version of $f(t)$, defined as:

$$f_1(t) = \begin{cases} 5e^{-2t}, & t \geq 0 \\ 0 & t < 0 \end{cases}. \tag{3-5}$$

In the context of our problem, such truncation will mean that the drug concentration is zero before the initial dose. The graphs of the truncated versions for $f_1(t)$ and $g_1(t) = f_1(t - 1)$ are given in Figure 3-2.

The analytic representation (3-5) has the disadvantage (with regard to the future use of the *BERKELEY MADONNA* syntax) that the function $f_1(t)$ cannot be written as a single algebraic expression. This can be corrected by using a *step function*. Consider the function:

$$\text{step}(h, a, t) = \begin{cases} h, & t \geq a \\ 0 & t < a \end{cases}, \tag{3-6}$$

whose graph is sketched in Figure 3-3. The function from Eq. (3-6) is called a step function with a step of height h at $t = a$.

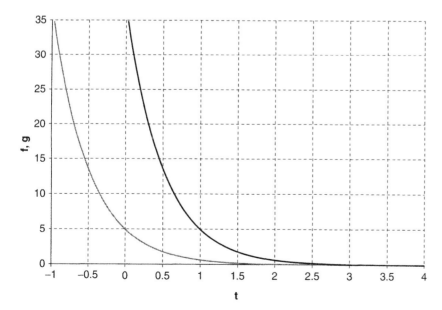

FIGURE 3-1.
Graphs of a function and its horizontal translation. Specifically, these are graphs of $f(t) = 5e^{-2t}$ (*gray line*) and $g(t) = f(t - 1) = 5e^{-2(t-1)}$ (*black line*).

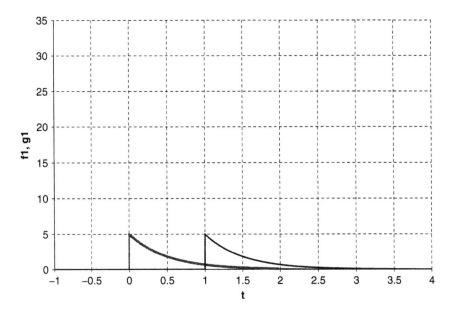

FIGURE 3-2.
Graphs of truncated functions. Specifically, graphs of $f_1(t)$ (*gray line*) and $g_1(t)$ (*black line*).

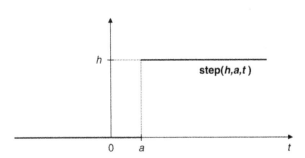

FIGURE 3-3.
Graph of a step function.

Using step functions, the functions $f_1(t)$ and $g_1(t)$ could be written as the following products of functions (see Exercise 3-1):

$$f_1(t) = f(t)\,\text{step}(1, 0, \text{t})$$

and

$$g_1(t) = g(t)\,\text{step}(1, 1, t). \tag{3-7}$$

SOFTWARE BACKGROUND

The software necessary for this project is *BERKELEY MADONNA*. It provides a built-in step function called step(h,a). Notice that the only difference in the syntax when compared with Eq. (3-6) is that the time variable t is not explicit.

In this laboratory exercise, we shall use various sums of functions and the process will be simplified by the use of *arrays*. Arrays represent collections of similar elements that differ from one another simply by an index that ranges between two fixed non-negative integer values.

For example, assume that we would like to consider the sum y of 20 different step functions of equal height $h = 1$ and $a = 1, a = 2, \ldots, a = 20$, respectively. One way to do this, of course, is to directly type the sum of all 20 step functions. If the number of step functions is 100 or, even worse, 1,000,000, this approach is certainly not feasible. The way to approach the problem is to define an array. In *BERKELEY MADONNA* this could be done by typing:

```
f[1..20] = step(1,i)
```

This statement establishes a collection of 20 step functions step(1,i), for which the index i takes all consecutive values from 1 to 20. The first element of the collection is $f[1] = $ step(1,1); the second one is $f[2] = $ step(1,2), and so on; and the last element is $f[20] = $ step(1,20).

If y denotes the sum of all elements in the collection (i.e., if $y = f[1] + f[2] + \ldots + f[19] + f[20]$), we can enter this in *BERKELEY MADONNA* by using the built-in function `arraysum`. This function returns the sum of the elements for the array given as its argument. Thus, if we enter in *BERKELEY MADONNA*

```
y = arraysum(f[*])
```

the sum $f[1] + f[2] + \ldots + f[19] + f[20]$ will be calculated and saved as a quantity denoted by y.

We stress that the default name for the time variable t in *BERKELEY MADONNA* is *time*, and in some of the exercises below we shall want to graph functions of t. This can be achieved by using the command

```
rename time = t
```

to replace the default time variable "time" with "t." If you choose not to do so, you will need to use *time* in place of t in all formulae that appear below.

DRUG DOSAGE PROBLEM

We are now ready to get back to our main problem. Recall that we assumed instantaneous and full absorption of the drug with each dose (i.e., the drug is administered intravenously, and thus the full amount enters the bloodstream) and exponential elimination, the mathematical form of which is given by Eq. (2-1). Equal doses of a drug should be administered at regular time intervals.

Note that when a dose is given (such as when you take an acetaminophen pill), it is given as an amount that may be expressed in terms of mass (such as 500 mg of acetaminophen) or in terms of an activity unit, such as 50 units of insulin. In the body, the dose is reflected as an increase in the concentration of the drug in the bloodstream. Concentration is an amount per unit volume and may be measured in such terms as ng/ml, mg/ml, or mg/L, because the drug is diluted in the blood (not to mention the fact that the drug will be distributed among other body compartments as well). Thus, we (and the drug manufacturers) are making an assumption about the volume of the typical patient. This "one dose fits all" approach seems to work reasonably well for adults. For children, however, the dosage will vary according to the child's weight, which is a reasonable proxy for volume. (Read the dosage information on a box of children's acetaminophen.) It should be noted that the dosage could also be calculated in terms of mg/kg total body weight.

Exercise 3-1

(a) Solve the equation in Eq. (2-1), assuming that a single dose $C(0) = C$ µg/ml is administered at $t = 0$ (and that the concentration of the drug in the bloodstream before this dose was negligible), to show that $C(t) = \begin{cases} Ce^{-rt} & t \geq 0 \\ 0 & t < 0 \end{cases}$. What is the concentration of the drug at time $t = 10$?

(b) Use a step function to write $C(t)$ as a single algebraic expression.

Exercise 3-2

(a) If a single dose C µg/ml is given at time $t = 2$, the result should be the same as in Exercise 5, except that the elimination process will begin at $t = 2$ instead of $t = 0$. In other words, there will be a shift in time. Give the formula for the concentration that corresponds to that in Exercise 1(a).

(b) Use a step function to write your answer from (a) as a single algebraic expression.

(c) If a single dose C µg/ml is given at time $t = i$, show that the residual concentration at any moment t is given by $C(t) = Ce^{-r(t-i)}$ step $(1,i)$.

Exercise 3-3

Assume doses of C µg/ml each are administered at times $t = T$, $t = 2T$, and $t = 3T$. Show that $C(t) = Ce^{-r(t-T)}$ step$(1,T) + Ce^{-r(t-2T)}$ step$(1,2T) + Ce^{-r(t-3T)}$ step$(1,3T)$.

Exercise 3-4

In *BERKELEY MADONNA*, define the following array:

```
rename time = t

f[ 1..20]  = C*exp(-r*(t-i*t_0))*step(1, i*t_0)
```

and define the sum of its elements as

```
y = arraysum(f[ *] )
```

Do not run the model yet. If C is the dose administered every $t\_0 = t_0$ hours and r is the elimination constant for the drug, what is the physiological meaning of the sum y?

Exercise 3-5

For this exercise, make sure that the autoformat feature for the y-axis is turned off and the scales on the left y-axis and right y-axis are the same. To do this, double-click anywhere on one of the axes. When a dialog window appears, uncheck the Auto checkboxes across from "Left Y Axis" and the "Right Y Axis" and enter the following ranges for both axes:

> Minimum: 0;
> Maximum: 40;
> # Div: 8

Click OK. More details on changing the axes settings can be found in Lab 4.

(a) Complete the model from Exercise 3-4 with the following values: $C = 4$, $p = 0.2$, $t_0 = 1$, and STOPTIME $= 40$. Run the model with these values.
    Remove the graph of f[] by disengaging the button in the lower left corner – it is unnecessary. Examine the graph of the concentration given by y. Include the graph of y with your laboratory report.

(b) What happens after time t = 20?"

(c) Define sliders on the parameters C, r, and t_0, and study the impact they have on the drug concentration when you change the parameter values one at a time. Summarize your observation in your report.

(d) Examine the peaks and the nadirs in the graph of y. What is causing the jumps?

(e) Reset the parameters C, r, and t_0 to their values from part (a). Does it appear that the sequence of nadirs stabilize around a specific value L? If so, give an estimate for L. Make a prediction about how your model will respond to increasing the number of administered doses while continuing to follow the established regimen.

Exercise 3-6

To take a closer look at the last question in Exercise 3-5, we introduce yet another parameter into our model. If n is the number of administered doses, modify your *BERKELEY MADONNA* code to allow examining the effect from changing n in the model. First, use $n = 20$ and $C = 4, r = 0.2$, and $t_0 = 1$ to verify that you will obtain the same graph as in Exercise 5(a). Next, use a slider for n to examine the effect on the concentration as n increases.

(a) Does it still appear that when n becomes large, the sequence of nadirs converges to L?

(b) If the sequence of nadirs converges to L, what does the sequence of peaks converge to? Visually estimate this value.

Save your model and close the Graphics window. The following example demonstrates how to calculate analytically the value L.

Assume that a certain drug with elimination constant r hours^{-1} is first administered at 1:00 A.M. at a dosage of C µg/ml, and in the same dosage every 3 hours afterwards. What is the concentration of the drug at 12:00 noon?

To answer this question, notice that by 12:00 noon we will have given four doses. The total concentration is the sum of the concentrations of each dose. Because the doses were given at different times, the effect at 12:00 noon is different for each dose. We give the data in Table 3-1.

The total concentration at 12:00 noon will be equal to the sum of all residual concentrations:

$$Ce^{-11r} + Ce^{-8r} + Ce^{-5r} + Ce^{-2r} = Ce^{-2r}(e^{-9r} + e^{-6r} + e^{-3r} + 1). \quad (3\text{-}8)$$

If we denote $b = e^{-3r}$, then $b^2 = e^{-6r}$, and $b^3 = e^{-9r}$. So with this notation, the expression (3-8) could be written more compactly as

$$Ce^{-2r}(b^3 + b^2 + b + 1).$$

Time of Dose	Length of Time in Body (hours) by 12:00 noon	Residual Concentration from the Dose at 12:00 noon [µg/ml]
1:00 A.M.	11	Ce^{-11r}
4:00 A.M.	8	Ce^{-8r}
7:00 A.M.	5	Ce^{-5r}
10:00 A.M.	2	Ce^{-2r}

TABLE 3-1.
Residual concentration of the drug in the blood as a function of the length of time since intake.

If we look closely at the example, we can extract a more general result. First, in the term e^{-3r}, the term 3 arises from the time between dosages. Notice also that the parenthetical expression contains four terms:

$$b^3 + b^2 + b + 1 = b^3 + b^2 + b + b^0$$

and that we have given four doses. The value C is the concentration, and the 2 in the factor Ce^{-2r} comes from the fact that it has been 2 hours from the last administered dose.

Exercise 3-7

Suppose that beginning at time $t = T$, we administer a dose of C μg/ml every T hours. Show that the drug concentration at time $t = (n + 1)T$ (just before the n-th dose is administered) because of the doses given at $t = T, t = 2T, \ldots,$ and $t = nT$ is

$$R_n = Ce^{-Tr}[(e^{-Tr})^{n-1} + (e^{-Tr})^{n-2} + (e^{-Tr})^{n-3} + \ldots + 1]. \qquad (3\text{-}9)$$

Thus, R_n are the concentrations corresponding to the nadirs on the graph from Exercise 3-5. Immediately after the next dose, the concentration rises to:

$$R_n + C = C[(e^{-Tr})^n + (e^{-Tr})^{n-1} + (e^{-Tr})^{n-2} + \ldots + e^{-Tr} + 1]. \qquad (3\text{-}10)$$

Exercise 3-8

Using Eqs. (3-2) and (3-4), show that $R_n = Ce^{-Tr}\dfrac{1 - (e^{-Tr})^n}{1 - e^{-Tr}}$.

Exercise 3-9

Using Exercise 8 and the result (3-3) with $a = Ce^{-Tr}$ and $b = e^{-Tr}$, show that

$$L = \lim_{n \to \infty} R_n = \frac{Ce^{-rT}}{1 - e^{-rt}} = \frac{C}{c^{rT} \quad 1}. \qquad (3\text{-}11)$$

Exercise 3-10

Is R_n larger or smaller than L? Explain why. What is the physiological meaning of L?

Exercise 3-11

Add the expressions for L and $H = L + C$ to your *BERKELEY MADONNA* model. To do this, enter

```
L = C*(1/(exp(r*t_0) − 1))

H = L + C
```

Remember that we are using $T = t_0$. Run the model. Use sliders for n, C, r, and t_0 to observe the effect caused by parameter changes. For large enough values of n, does it appear that the concentration is between the levels L and $H = L + C$? Save your model and close the Graphics window.

We are now ready to give a solution to our main problem. Knowing the MEC, the MTC, and the half-life of the drug (or its elimination constant r), we would like to design a safe therapeutic regimen with maximal benefits. Equal doses C of the drug should be given at equal time intervals T. Once the concentration reaches the minimal effective level, it should remain between the MEC and the MTC levels (the therapeutic window).

We saw in Exercises 3-5, 3-6, and 3-11 that, after a few doses, the concentration is almost between L and $H = L + C$. Because one of our goals is to maintain the concentration between the minimal effective and the minimal toxic levels, we would like to have

$$L = \text{MEC}, \quad H = L + C = \text{MTC}. \tag{3-12}$$

Exercise 3-12
....

When the MEC and MTC are known, what single dose C would ensure that after a certain number of doses the concentration will remain between the MEC and MTC?

Exercise 3-13
....

Using the value of $L = \text{MEC}$ from Eq. (3-11) and the value of C from Exercise 3-12, show that the time between the doses T should satisfy the condition

$$MEC = \frac{MTC - MEC}{e^{Tr} - 1}.$$

Solve this equation for T to show that the time between doses T should be determined as

$$T = \frac{1}{r} \ln \frac{MTC}{MEC}.$$

Exercise 3-14
....

(a) Let MTC $= 19$ µg/ml and MEC$=14$ µg/ml. If $r = 0.5$ hours^{-1}, determine the amount of each dose and the time between doses.

(b) Using that the natural logarithm of x is entered in *BERKELEY MADONNA* as `logn(x)`, modify your model from Exercise 3-11 to include MTC and MEC as parameters, and remove C and t_0 from the parameter list (their values should now be expressed as functions of MTC and MEC). If $n = 20$ and $r = 0.5$ hours^{-1}, find how many doses of the drug will need to be administered for the concentration to enter and remain in the therapeutic window.

The restriction that all doses have to be the same has the obvious disadvantage that a certain build-up period is required before the concentration reaches the MEC level. For some drugs, such as certain antidepressants, a slow build-up period is necessary to minimize side effects. For many other common drugs, however, the dosage schedule tolerates a larger first dose aimed at achieving the MEC as quickly as possible. We illustrate this now.

Exercise 3-15

If the concentrations MEC, MTC, and the elimination constant r are as in Exercise 3-14, determine a drug intake schedule that maximizes the therapeutic effect of the drug under the following constraints:

1. All doses must be given at equal time intervals.

2. All doses, except possibly for the first dose, must be equal.

3. The maximal effective concentration must be achieved as quickly as possible.

4. The concentration of the drug should remain between the minimum effective and the minimum toxic levels at all times.

Hint: To the model from Exercise 3-14, add a term $A*\exp(-r*t)*\text{step}(1,0)$. This term represents a dose of A µg/ml administered at time $t = 0$. Choose different values of A until an (approximately) optimal value is obtained. Would it have been possible to make an educated guess for the value of A you found? Explain.

SOME CAVEATS

In our model, we have assumed that any of the drug taken is immediately assimilated into the bloodstream. This is not the case (unless the drug is given intravenously, in which case it is not too far from being factual). Adjustments will need to be made to the model if the drug is administered by a different route, such as orally, intramuscularly, or transdermally.

Exercise 3-16

Without getting into any mathematical details, describe what aspects of the model would change if the drug is not absorbed instantaneously (e.g., in the case of oral, intramuscular, or transdermal intake).

We also assumed that the drug should be taken at regular time intervals. As already mentioned, this is done for the purpose of convenience and is followed as a guideline when determining the dosage schedules for a large number of drugs. In emergency situations, drugs may be given intravenously and *continuously* in order to keep their concentration as close as possible to or at the MTC at all times.

We have also used the approximation that $R_n = L = \text{MEC}$ for large values of n. In fact, R_n is always slightly smaller than L, and there will be very small periods of time where the concentration is slightly below MEC. However, these times and differences are so small they can normally be ignored.

Exercise 3-17

What would one do if it were imperative that the concentration eventually be kept larger than MEC (i.e., if it were crucial not to ignore the small differences mentioned above)?

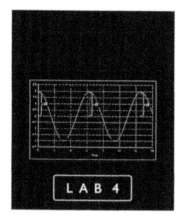

LAB 4

Epidemic Models

In this laboratory exercise, we focus on mathematical models that describe the spread of an epidemic. We shall examine continuous mathematical models of fixed populations, models in which the population is not fixed (e.g., the models in which we allow for births and deaths, or immigrations and emigrations), and models with delay. We generalize the concepts of equilibrium states and long-term behavior to multiple component systems and examine them both analytically and through time plots and phase diagrams.

In this lab, you will:

- Study epidemic models with regard to their underlying assumptions and the meaning of the model parameters.

- Understand the meaning of equilibrium states.

- Find equilibrium states for a dynamic model involving multiple groups.

- Derive information regarding equilibrium states from the model's phase trajectories.

- Practice conversion from time trajectories to phase trajectories and vice versa.

- Observe models that exhibit oscillatory behavior.

BIOLOGICAL BACKGROUND

We are all familiar with the progression of an infectious illness, from exposure through the miserable symptoms to recovery. A mother wipes her infant's runny nose and inadvertently rubs her own nose while on her way to the sink to wash her hands. The viruses on her hand adhere to mucous membrane cells, and an infection is initiated.

In the beginning, the host may not realize that she is infected. Following infection, the host enters a phase called the *incubation period*, during which the pathogen begins to multiply. The infected host shows no symptoms at all during this period, which usually lasts from several days to several weeks but may be as short as a few hours or as long as 10 or 12 years. The length of the incubation period is characteristic of the infectious organism, within limits determined by the health of the host and the route of infection. The healthier the host and the longer the path the infecting organism must travel, the longer the incubation period.

The host begins to feel ill during the *prodromal period* or *prodromium*. This is a short period of mild symptoms, which may be difficult to characterize as anything other than "not feeling quite right." The prodromium is rapidly followed by the *period of illness* or *period of invasion*, which is characterized by the most rapid reproduction of the pathogen. The host develops unmistakable symptoms that may be quite severe. If the disease is serious and the immune response is weak, delayed, or absent, the host may die during this phase. The peak of this phase is called the *fastigium*.

If all goes well, the immune system will begin to bring the infection under control, and the infection will enter the *period of decline*. The host will begin to get better and enter the *convalescent period*. With many diseases, the host will acquire long-lasting resistance to the disease, such that when the pathogen is encountered again, its immune system will mount such a rapid and strong reaction that an infection will not be established. This phenomenon is described by the term *immunity*.

MATHEMATICAL BACKGROUND

The models we consider in this laboratory exercise are given by a special type of systems of differential equations—each equation in the system represents the rate of change in time of a single quantity. Values of the model variables for which *all* rates of change in the model are equal to zero, give the *equilibrium states* for the model. The *long-term behavior* of the model variables is established in terms of their limits when t→∞.

For example, in a model with two quantities x and y given by the equations

$$\frac{dx}{dt} = f(x, y)$$

$$\frac{dy}{dt} = g(x, y),$$

the equilibrium states are points (x_0, y_0) where $f(x_0, y_0) = 0$ and $g(x_0, y_0) = 0$, and the long-term behavior will be derived by considering the limits $\lim_{t \to \infty} x(t)$ and $\lim_{t \to \infty} y(t)$. Depending upon the existence of specific values of these limits, the solutions may converge to equilibrium or not, and in the latter case, it may oscillate.

The behavior of the model solutions near equilibrium points could serve as a basis for classifying the equilibrium points into different groups. This is best done in considering the *phase trajectories* of the model. Those are obtained by plotting one of the model variables versus another; for the example above, the phase plot will be a plot of x versus y. Every value of the time variable t determines one point $(x(t), y(t))$ of the phase trajectory. The curve resulting from plotting the points $(x(t), y(t))$ for all values of t generates the (x,y)—phase trajectory. If the model uses more than two variables, the specific choice of variables for the phase plot depends on the questions of interest. For example, when modeling the spread of an epidemic, the graph that presents the number of people who have the disease and can infect others (the infectives) against the number of people who do not have the disease and can be infected (the susceptibles) may be of interest.

For us, phase diagrams will be crucial as a tool for determining the long-term behavior of model solutions and particularly for examining their behavior in close proximity to the equilibrium states. An equilibrium point (x_0, y_0) of the system of Eq. (2-8) is called *stable* if for any region in the plane U that contains (x_0, y_0), there exists a smaller region V contained in U, such that all trajectories that initiate from V remain in U for all $t > 0$. Intuitively, all trajectories that initiate sufficiently close to a stable equilibrium point "remain close" to that point for any $t > 0$. An equilibrium point that is not stable is called *unstable*. The analytic conditions for determining the stability of an equilibrium point are more technical and can be found in Chapter 2, Section VI.3 in the textbook.

SOFTWARE BACKGROUND

The software necessary for this lab is *BERKELEY MADONNA*. Some of the software features not utilized in previous laboratories are introduced here.

Phase Plots. We briefly describe the *BERKELEY MADONNA* commands necessary to obtain a phase plot of the model variables. We shall use the following example as an illustration. This model will be considered in more detail in Exercise 14.

$$\frac{dS}{dt} = -(0.00002)IS + 1$$

$$\frac{dI}{dt} = (0.00002)IS - 0.04I. \tag{4-1}$$

Exercise 4-1

(a) Enter the model:

```
STARTTIME = 0

STOPTIME = 10000

DT = 0.02

d/dt(S) = −0.00002*S*I + 1
```

```
d/dt(I) = 0.00002*S*I - 0.04*I

init S = 2500

init I = 15
```

Run the model to obtain the numerical solution and time plots of the variables S (susceptibles) and I (infectives) shown in Figure 4-1.

(b) *Changing the axis settings.* By default, BERKELEY MADONNA uses autoformatting of the graph axes. The range and the grids of the axes are being chosen by the program in a way that provides for optimal display of the model's solutions. It is also important to know that the program often utilizes both a left y-axis and a right y-axis when systems with multiple components are considered. This means that different variables could be plotted on these axes *using different grids.* For example, in Figure 4-1, the S variable is plotted on the left y-axis while the I variable is plotted on the right y-axis. If standardizing the output is desired, the autoformat feature can be turned off. To do this, follow the steps below:

1. Double-click anywhere on one of the axes.

2. A dialog window as in Figure 4-2 will appear. Depressing the Legend button located across from the Run button in the graphical window (with an icon that features the letter L) will allow you to see the graph legend. In the same way, activating the Parameter button (its icon features the letter P) will allow you to see the values of the parameters used to obtain the graph.

3. Uncheck the Auto checkboxes. This will allow you to enter ranges for the x- and y-axes individually as well as to specify the grid on each of the axes by giving the number of division points.

4. Enter the desired information. For this exercise, we want to make the grids on the left y-axis and the right y-axis be the

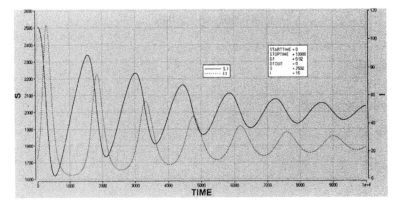

FIGURE 4-1.
Time plots of S and I for the model given in Exercise 4-1(a).

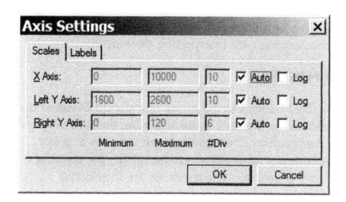

FIGURE 4-2.
BERKELEY MADONNA Axis Settings dialog box.

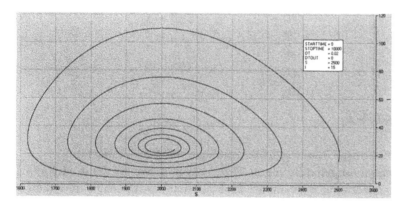

FIGURE 4-3.
Phase plot of S versus *I*.

same: minimal value of 0 and maximal value of 2600, with 13 division points for the whole range from 0 to 2600. Click OK.

5. To turn the autoformatting back on, double click on either axis and check all of the Auto checkboxes.

(c) *Phase plots.* To obtain the phase trajectory graph of *S* versus *I*, select Graph > Choose Variables from the main menu. Select S for the X Axis variable, then highlight S in the Y Axis text area and click Remove to remove it from the list. Click OK. You should now have the phase plot of *S* versus *I*, as in Figure 4-3.

(d) *Initial conditions. BERKELEY MADONNA* allows for selecting a point (S(0), I(0)) in the (S,I) plane that corresponds to the initial conditions of the dynamical system by simply clicking at the desired point. To be able to do this, make sure that the I_c button from the *BERKELEY MADONNA* button bar is activated. Each click on the (S,I) plane now generates a trajectory that corresponds to the initial condition defined by the location of your click. Make several clicks to assign new initial conditions and become familiar with this feature of the software.

Note: It may be useful to turn the autoformat of the axes feature off at first until you feel comfortable with interpreting the output.

PART I: THE SIR MODEL

Prior to infection, the host belongs to the group of individuals called the *susceptibles*, those who can be infected. After infection, the host belongs to the group called the *infectives*. These individuals are able to infect other susceptible individuals. When they recover or die, the infectives enter the group called the *removals*. The removals are considered immune and cannot be infected again.

The model is:

$$\frac{dS}{dt} = -kIS \qquad S(0) = S_0$$

$$\frac{dI}{dt} = kIS - aI \qquad I(0) = I_0 \qquad \qquad (4\text{-}2)$$

$$\frac{dR}{dt} = aI \qquad R(0) = 0$$

Exercise 4-2

(a) What major (medical) characteristics should the infectious disease modeled by Eq. (4-2) possess in order for the model to be considered reasonable? That is, what are the major hypotheses upon which the model from Eq. (4-2) is built?

(b) Explain the meaning of the parameters k and a in the context of this model.

(c) Notice that $\frac{dS}{dt} + \frac{dI}{dt} + \frac{dR}{dt} = \frac{d}{dt}(S + I + R) = 0$. Give the meaning of this fact in the context of the model.

Exercise 4-3

Enter the model given by Eq. (4-2) into *BERKELEY MADONNA* by typing the appropriate differential equations. Use the following parameters to run the model:

STARTTIME = 0

STOPTIME = 100

DT = 0.02

init S = 1000

init I = 1

init R = 0

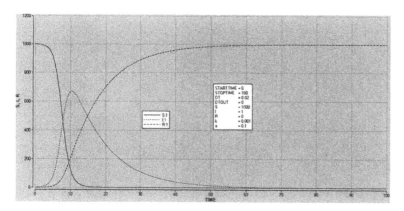

FIGURE 4-4.
Graphs of S, I, and R.

k = 0.001

a = 0.1

Your graph should be similar to the graph in Figure 4-4. If some of the trajectories are not displayed, make sure to depress the button in the lower left corner of the window frame labeled with the respective variable name. Comment on the medical meaning of the shapes of the graphs $S(t)$, $I(t)$, and $R(t)$.

Exercise 4-4

Comment on how you expect the graphs of $S(t)$, $I(t)$, and $R(t)$ to change if you increase or decrease the parameters k while the parameter a remains fixed. Justify. Next, comment on how you expect the graphs of $S(t)$, $I(t)$, and $R(t)$ to change if you increase or decrease the parameters a while the parameter k remains fixed. (You are not allowed to use *BERKELEY MADONNA* for this exercise!)

Exercise 4-5

Now, define sliders[1] for k and a, and use *BERKELEY MADONNA* to examine whether your predictions from Exercise 4 were correct. If they were not, reconsider your answer and make sure that you understand the reasons. Repeat until you feel comfortable with predicting the change in graphs resulting from changing the values of k and a.

Although the SIR model is a relatively simple mathematical model, it is not possible to solve this system of differential equations analytically. Because $S + I + R = N =$ const, where N is the total number of individuals in the system under consideration, knowing two out of the three quantities S, I, or R is enough. (Why?) We begin with looking at the trajectories of S (susceptibles) versus I (infectives).

---

1. Revisit Exercise 1–5 from Lab 1 if you need a refresher on how to do this.

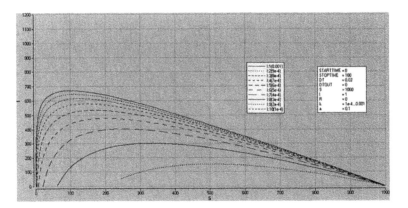

FIGURE 4-5.
Effects of changing a parameter value. Result of a batch run with varying values of *k*.

Exercise 4-6

The graph in Figure 4-5 was obtained using *BERKELEY MADONNA* by performing a batch run with 10 different values of the parameter *k*—from *k* = 0.0001 to *k* = 0.001, with increments of 0.0001. Each trajectory represents a different value for the parameter *k*. Examine Figure 4-5 and, based on the medical meaning of the parameter *k*, determine which one of the trajectories corresponds to the largest value of *k*. Which one corresponds to the lowest value of *k*. What considerations did you use? Explain.

Exercise 4-7

(a) In Figure 4-5, indicate the direction of change with time for the phase trajectories by inserting arrows along each of the graphs.

(b) For the graphs in Figure 4-5, interpret the meaning of the points where the trajectory intersects the horizontal axis in the context of the model.

(c) According to the SIR model, will the disease die out in a long run or will it remain endemic within the population? Explain your answer.

Exercise 4-8

Use *BERKELEY MADONNA* to investigate how changing the values of the parameters *k* and *a* affects the trajectories of *S* versus *I*. You may use sliders and/or batch runs[2], as you find appropriate.

---

2. Review Exercise 4, Lab 1 if you need a refresher on how to use batch run in *BERKELEY MADONNA*.

The most important question about a potential epidemic is whether the infection is going to spread, and if so, how fast. It is also important to know when the infection will start to decline. The answers to these questions depend on the specific characteristics of the infectious disease (represented in the model by the parameters k and a) and on the initial conditions S_0 and I_0.

The spread of the disease is measured by the number of infectives $I(t)$. We say that an *epidemic occurs* when the number of infected individuals $I(t)$ increases as time goes by. If this number does not increase, there is no epidemic. When $I(t)$ decreases to zero as time goes by, we say that the disease dies out.

Next, we attempt to find a condition for the model from Eq. (4-2) that will allow us to determine whether or not there will be an epidemic.

Exercise 4-9

(a) For Eqs. (4-2), study the signs of the derivatives $\dfrac{dS}{dt}, \dfrac{dI}{dt}$, and $\dfrac{dR}{dt}$. Which derivatives will always be positive or negative, regardless of the values of $S(t)$, $I(t)$, and $R(t)$?

(b) What derivative may be changing sign?

(c) Consider the derivative $\dfrac{dI}{dt}$. For $t = 0$, we calculate

$$\frac{dI}{dt}(0) = kS(0)I(0) - aI(0) = (kS(0) - a)I(0) = (kS_0 - a)I_0.$$

Determine a condition that S_0 has to satisfy in order to have $\dfrac{dI}{dt} > 0$ at $t = 0$. What is the condition that S_0 has to satisfy in order to have $\dfrac{dI}{dt} < 0$ at $t = 0$?

(d) Show that Eqs. (4-2) imply that $S(t) < S_0$ for any value of $t > 0$.

(e) Using your answers from parts (c) and (d), show that if $S_0 < \dfrac{a}{k}$, then $\dfrac{dI}{dt} < 0$ for all $t \geq 0$. What does this imply about the disease—is there an epidemic or is the disease dying out?

(f) Using your answers from parts (c) and (d) show that if for some value $t \geq 0$, $S(t) > \dfrac{a}{k}$, then $\dfrac{dI}{dt} > 0$. What does this imply about the disease—is there an epidemic or is the disease dying out?

This exercise demonstrates that a specific *threshold value* for the initial size of the susceptible group is of critical importance to the SIR model

(4-2). You found that if $S_0 < \frac{a}{k}$, there is no epidemic, whereas if $S_0 > \frac{a}{k}$, there is an epidemic. The threshold value $r = \frac{a}{k}$ is called the *relative removal rate*. Its reciprocal $\frac{k}{a}$ is the infection's *contact rate*. The number $\sigma = \frac{S_0 k}{a}$ is the basic *reproduction number* for the infection. It represents the average number of infections that a single infective will produce in a wholly susceptible population.

Exercise 4-10

Run the model in *BERKELEY MADONNA* again, and consider the graph of the trajectory S versus I. By only looking at the plot (and without calculating the relative removal rate explicitly!), determine the point on the horizontal axis for the relative removal rate $r = \frac{a}{k}$. Based on Exercise 4-9 (f), how can you recognize this point?

Exercise 4-11

For the initial conditions and values for the parameters given in Exercise 3, calculate the relative removal rate, the contact rate, and the basic reproduction number for the infection.

Exercise 4-12

Turn off the autoformatting of the axes, and activate the I_c button located above the graph. A click on the plot will now generate a trajectory with initial conditions that are determined by the coordinates of the point at which the click occurs. Make several mouse clicks that correspond to initial conditions with S_0 smaller than and larger than the value of relative removal rate r. What is the difference in the trajectories? How does the difference in behavior relate to the presence or lack of an epidemic?

Exercise 4-13

What does the magnitude of the relative removal rate $r = \frac{a}{k}$ mean in terms of the severity of the epidemic? More specifically,

(a) If r is small, do we have a mild or a severe epidemic?

(b) If r is large, do we have a mild or a severe epidemic?

PART II: OTHER EPIDEMIC MODELS

For many infectious diseases, periods of acute epidemic outbreaks may be separated by relatively quiet periods. A mathematical model capable of describing such events should therefore have oscillating time trajectories for S and I. In this section, we consider epidemic models that have such solutions. The oscillations are about equilibrium values, and the type of the equilibrium state determines the type of oscillations.

Exercise 4-14

We now consider a generalization of the model (4-1) given by the equations

$$\frac{dS}{dt} = -kIS + b$$
$$\frac{dI}{dt} = kIS - aI. \tag{4-3}$$

This model differs from the SIR model in its assumption that the total population does not remain constant. For this model, answer the following questions:

(a) The model assumes a constant rate at which new susceptibles enter the system. Identity the model parameter that describes this behavior.

(b) The equations of model (4-3) reflect a hypothesis with regard to the possibility of repeated infections. What is this hypothesis? According to the model described by Eq. (4-3), would it be possible to be infected again upon recovery?

Exercise 4-15

(a) Find the equilibrium states for the model described by Eq. (4-3), expressed in terms of the model parameters a, b, and k.

(b) Give the model equilibrium states for $a = 0.04$, $b = 1$, and $k = 0.00002$.

(c) Enter the Eqs. (4-3) into *BERKELEY MADONNA*, and run it with initial conditions $S(0) = 2500$, $I(0) = 15$, STARTTIME $= 0$, STOPTIME $= 50000$, DT $= 0.02$ and the parameter values from part (b) to observe the solution. Consider the phase plot of S versus I as in Exercise 1. Describe the behavior of the phase trajectory near the equilibrium state.

(d) Is there convergence to equilibrium as $t \rightarrow \infty$? If so, how is such convergence observed in the: (i) time trajectories? (ii) phase trajectories?

(e) Use the I_c button to generate phase plots corresponding to different initial conditions. Does the behavior of the phase trajectory appear to be similar or different? Does it appear to you that this behavior depends at all on the initial conditions of the model?

(f) Based on your answers for parts (c) and (d), classify the equilibrium state(s) from part (b) as stable, asymptotically stable, or unstable.

(g) Use the analytic test for stability introduced in Chapter 2, Section VI.3 of the text to confirm your answer for part (f).

(h) For a disease described by Eq. model (4-3), answer the following question: Will the disease die out, or will it become endemic within the population?

Consider now the following model similar to that described by Eq. (4-3). In this modified version, we allow for births to the susceptibles and consider the infectives to be removed by immunity. Thus, as in the model described by Eq. (4-3), the population does not remain constant. The differential equations are

$$\frac{dS}{dt} = -kSI + cS$$

$$\frac{dI}{dt} = kSI - aI,$$

(4-4)

where $c > 0$ is the per capita birth rate and a is the per capita rate of recovery. As before, the parameter k is a measure of how contagious the disease is.

Exercise 4-16
....

Enter Eq. (4-4) in *BERKELEY MADONNA*, and run it with values $k = 0.00003$, $a = 0.03$, $c = 0.015$ and initial conditions as in Exercise 1: $S(0) = 2500$, $I(0) = 15$. Run the model. Define sliders on the parameter values, and adjust those values in a way that will allow you to observe that the solutions exhibit steady oscillations (similar to those shown in Figure 4-6).

(a) Include the graph with your lab report, and specify the values of the parameters for which the graph was obtained.

(b) For the same values of the parameters, examine the phase trajectory of the model. What does it look like? Explain why.

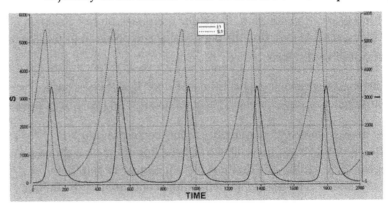

FIGURE 4-6.
Graph of a model giving oscillating time trajectories for S and I.

(c) Compute the equilibrium states for this model for $k = 0.00002$, $a = 0.04$, $c = 0.01$. Next, in the phase plot of the solution, engage the I_c button and experiment with different initial conditions close to each of the equilibrium states. According to the definition of stability given in the beginning of this laboratory exercise, would you classify each of them as stable, asymptotically stable, or unstable? Justify your answer.

All of the models we used so far were built on the assumption that people who recover from the disease become permanently immune and cannot contract the disease again. Although for certain diseases (e.g., chicken pox) this is a viable assumption, there are other diseases (e.g., brucellosis or the sexually transmitted diseases gonorrhea, chlamydia, and syphilis) for which only limited immunity is received from having the disease. This immunity is lost after some time has elapsed, and the length of this time interval depends on the disease. To model this situation, we next consider an SIR model with delay.

As a first approximation, we could assume that the length of this temporary immunity is constant—those who have recovered from the disease lose immunity after a fixed time D. Thus, all removals remain in the R group for a fixed time D before joining the group of the susceptibles. Accordingly, at any time instance t, the flow from R into S will be at a rate equal to the flow from I into R at time $t - D$. From the SIR model, the latter is given by the term $aI(t - D)$, and this leads to the model from Eq. 4-5 below.

$$\frac{dS}{dt} = -kS(t)I(t) + aI(t - D)$$

$$\frac{dI}{dt} = kS(t)I(t) - aI(t) \qquad (4\text{-}5)$$

$$\frac{dR}{dt} = aI(t) - aI(t - D).$$

Entering Models with Delay. BERKELEY MADONNA provides a built-in function for incorporating delay in the models. This is the function

```
delay(x,d)
```

that represents the value of x delayed by time d. Thus, the model given by the Eqs. (4-5) can be entered as

```
d/dt(S) = -k*S*I + a*delay(I,D)
```

```
d/dt(I) = k*S*I - a*I
```

```
d/dt(R) = a*I - a*delay(I,D)
```

Note that in simulations involving a delay $D > 0$, initial conditions for the delayed functions must be provided for an entire interval of length D.

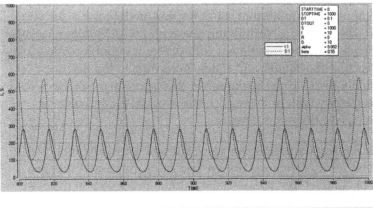

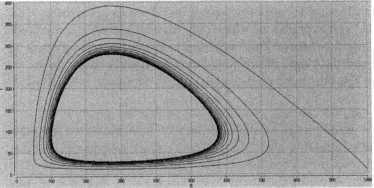

FIGURE 4-7.
Periodic solution for the SRI model with delay. Panel A: Time trajectories of $S(t)$ and $I(t)$ for $800 \leq t \leq 1000$; panel B: Phase plot of I vs. S.

By default, *BERKELEY MADONNA* extends the specified initial condition over the interval $[-D, 0]$. In this example, this menas that $I(t) = I(0)$ for t.

Exercise 4-17

An interesting feature of this model is that, for certain values of the parameters and the initial conditions, a convergence to a steady periodic cycle is possible. Enter the model into *BERKELEY MADONNA* using the following initial conditions: $S(0) = 1000$, $I(0) = 100$, and $R(0) = 0$. Use STOPTIME = 500, DT = 0.02, and initial values for the parameters of your choice (for example, $a = 0.5, k = 0.001, D = 9$ would be a reasonable initial combination).

(a) Define sliders on k, a, and D, and explore possible combinations of values while attempting to obtain an oscillating solution similar to the time and phase trajectories given in Figure 4-7. Record the values of the parameters that produce such a solution.

(b) Are there values for the parameters for which the model exhibits a convergence to a stable equilibrium? If so, record the parameter values for which this occurs.

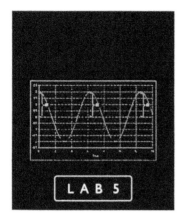

LAB 5

Predator–Prey Models

In this laboratory module, we examine mathematical models that describe the behavior of two populations that interact with one another. Because of this interaction, the sizes of the populations are dependent upon one another.

More specifically, in this laboratory exercise, you will explore the following concepts:

- Use systems of differential equations to describe the dynamics of the population sizes.

- Derive information about the dependency between population sizes and the equilibrium states of the system from the time trajectories and the phase trajectories of the model solutions.

- Understand the relationship between time and phase trajectories and how to obtain one of the trajectories from the other.

- Understand the behavior of the model phase trajectories in a neighborhood of stable equilibrium points and in a neighborhood of a repellor.

- Understand that the type of an equilibrium point may change with a change of the values of the model parameters.

BIOLOGICAL BACKGROUND

When two or more different species interact, the interaction could take many different forms. One population may feed upon another, or two may compete with each other to feed upon a third. Predation may be found among organisms both large and small—there are predatory insects that prey upon other insects, as well as predatory mammals whose prey may be smaller mammals. Competition may occur over any valuable aspect of the habitat, including nesting sites or places to hide. This competition may be classified as *exploitative* or *resource competition*, where the two species use the same resources, or as *interference competition*, where the two species cause harm to each other in the process. Associations between species need not always be harmful. *Mutualism*, which is also sometimes called *symbiosis*, is a state in which two species live in a close association with each other and in which both benefit in the process. Lichens are a good example of this kind of association, where an alga and a fungus combine to wrest a living from some extremely inhospitable environments, such as a bare rock face or a tree trunk.

One manner in which two populations of organisms may interact is *predation*, in which one type of organism is food for another. If the food organism is a plant, then the relationship is described as *herbivory*, which may be seen as a subset of predation. In the context of predation, the following question is important: What would happen to the sizes of populations when one species preys upon another?

One might conjecture that several different outcomes are possible. The predator may eliminate the prey, and, unless it finds something else to eat, the predator, would then follow its prey into extinction. The prey may evade the predator, and the predator would then starve. Finally, the predator and prey may exist in a balance, with each population exerting some control over the other in a manner that sustains both populations.

In this laboratory exercise, we explore mathematical models of predator–prey interactions in an attempt to provide answers to these questions.

MATHEMATICAL BACKGROUND

As in the previous laboratory project (4), the models that we consider here will be given by systems of differential equations where each equation in the system represents the rate of change in time of a single quantity. If a model with two quantities x and y is defined by the equations

$$\frac{dx}{dt} = f(x, y)$$
$$\frac{dy}{dt} = g(x, y),$$

(5-1)

the *null clines* of the model are determined by the conditions $\frac{dx}{dt} = 0$ or $\frac{dy}{dt} = 0$. The graphs of the null clines are curves in the plane corresponding to the graphs of $f(x, y) = 0$ and $g(x, y) = 0$ in the (x, y) plane. The *x-null cline* is determined by the condition $f(x, y) = 0$, and the *y-null cline* is determined by the condition $g(x, y) = 0$.

Values of the model variables for which *all* rates of change in the model are equal to zero give the *equilibrium states* for the model. The equilibrium states can, therefore, be obtained as the intersection points of the null clines.

The *long-term behavior* of the model variables is established in terms of their limits when t→∞. Thus, for the system (5-1) the equilibrium states are points (x_0, y_0), where $f(x_0, y_0) = 0$, and $g(x_0, y_0) = 0$, and the long-term behavior will be derived by considering the limits $\lim_{t\to\infty} x(t)$ and $\lim_{t\to\infty} y(t)$. Depending upon the existence or specific values of these limits, the solutions may converge to equilibrium or not and, in the latter case, it may oscillate.

An equilibrium point (x_0, y_0) of the system of Eqs. (5-1) is called *stable* if for any region in the plane U that contains (x_0, y_0), there exists a smaller region V contained in U, such that all trajectories that initiate from V remain in U for all $t > 0$. Intuitively, all trajectories that initiate sufficiently close to a stable equilibrium point "remain close" to that point for any $t > 0$. An equilibrium point that is not stable is called *unstable*. A subclass of stable points is of special importance. An equilibrium point (x_0, y_0) of the system of Eqs. (5-1) is called *asymptotically stable* when *all* trajectories that start in some region that contains the point (x_0, y_0) converge to the point (x_0, y_0) as t becomes large. A stable equilibrium point (x_0, y_0) that is not asymptotically stable is called *neutrally stable.*

The analytic condition for determining the stability of an equilibrium point can be found in Chapter 2, Section VI.

In this laboratory module, we continue to explore the behavior of solution trajectories in close proximity to the model's equilibrium states by using both phase plots and time plots for the model variables. We also examine a special type of unstable equilibriums called *repellors* defined by the condition that any trajectory that comes close to a repellor is forced away from it. More details, including the analytical definition of a repellor and an important result regarding the possible existence of periodic solutions in a neighborhood of a repellor (Poincaré–Bendixson's Theorem) can be found in Chapter 2, Section VI.

SOFTWARE BACKGROUND

The software necessary for this lab is *BERKELEY MADONNA*. No new features of the software will be required for this laboratory module. You will need to be comfortable with obtaining and interpreting time and phase plots of the model solution, defining and using sliders for the model parameters, changing the model's initial conditions, changing the axis settings of the plots, and other basic *BERKELEY MADONNA* features. For more details regarding these features, refer to the earlier laboratory projects.

PART I: THE MATHEMATICAL MODEL OF LOTKA–VOLTERRA

In the 1920s, two independent researchers, A. Lotka and V. Volterra, developed the first mathematical model of predator–prey interaction. This has come to be known as the Lotka–Volterra model.

The hypothesis is that we have two species: a predator (say, owls, O) and a prey (voles, V). The owls have voles as a major source of food. In the Lotka–Volterra model, if there are no owls then the voles will grow at a constant per capita rate. From our earlier work, we know that it is unrealistic for any species to grow indefinitely at a constant per capita rate, and this is a major weakness of the model.

Our earlier experience leads us to believe that the rate of growth of a species will be the net per capita growth rate of the species multiplied by the number of individuals. It is reasonable that as the food supply for the owls (i.e., the voles) increases, the net per capita rate of growth of the owls will increase, but, with more owls to eat the voles, the population growth of the voles will diminish. In other words, the growth rate of each species depends on the amount of the other species.

Thus, we have

$$\frac{dV}{dt} = r_1(O)V$$

$$\frac{dO}{dt} = r_2(V)O,$$

where $r_1(O)$ is the net per capita growth rate of the voles (which depends on the number of the owls) and similarly for $r_2(V)$. Our problem now is to form equations for $r_1(O)$ and $r_2(V)$.

For $r_1(O)$, we hypothesize that without any voles the owls will die out at a constant per capita rate, but, as the vole population increases, the owls will have a food supply and have the opportunity to grow. Thus,

$$r_1(O) = \alpha - \gamma O,$$

where α is the per capita growth rate of the voles in the absence of owls and γ is the per capita death rate of the voles per owl.

So we have $\dfrac{dV}{dt} = (\alpha - \gamma O)V = \alpha V - \gamma OV.$

Exercise 5-1

In the Lotka–Volterra models, the growth rate of the voles is given by

$$r_2(V) = -\delta + \varepsilon V,$$

so that

$$\frac{dO}{dt} = (-\delta + \varepsilon V)O.$$

Explain the meaning of the parameters δ and ε.

Exercise 5-2

Consider the following Lotka–Volterra model with $\alpha = 0.1$, $\gamma = 0.004$, $\delta = 0.02$, and $\varepsilon = 0.0001$:

$$\frac{dV}{dt} = (0.1 - 0.004O)V$$

$$\frac{dO}{dt} = (-0.02 + 0.0001\ V)O.$$

(a) For what values of O and V will $\dfrac{dO}{dt} = 0$? What does this say about the change in the population of owls?

(b) For what values of O and V will $\dfrac{dO}{dt} = 0$ and $\dfrac{dV}{dt} = 0$? What will happen in this case?

(c) Give the region in the (V,O) plane where $\dfrac{dO}{dt} > 0$. Give the region in the (V,O) plane where $\dfrac{dO}{dt} < 0$.

(d) Give the region in the (V,O) plane where $\dfrac{dV}{dt} < 0$. Repeat with $\dfrac{dV}{dt} > 0$.

In the (V,O) plane, if $\dfrac{dO}{dt} > 0$, then O is increasing. If $\dfrac{dV}{dt} < 0$, then V is decreasing. Thus, for each nonequilibrium point in the (V,O) plane, we have associated an arrow. If, for example, $\dfrac{dO}{dt} > 0$, then the arrow will point upward; if $\dfrac{dV}{dt} < 0$, then the arrow will point to the left. This combination, where $\dfrac{dO}{dt} > 0$ and $\dfrac{dV}{dt} < 0$, would give an arrow such as in Figure 5-1.

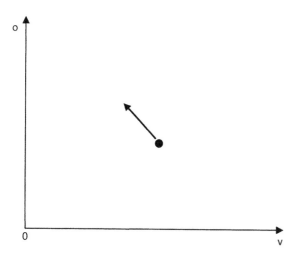

FIGURE 5-1.
Trajectory for $\dfrac{dO}{dt} > 0$ and $\dfrac{dV}{dt} < 0$. The direction of movement is indicated by an arrow.

Exercise 5-3

(a) Divide the first quadrant in the (V,O) plane into four regions determined by the null clines $\frac{dV}{dt} = 0$ and $\frac{dO}{dt} = 0$.

(b) In each of the four regions, determine the sign of $\frac{dV}{dt}$ and $\frac{dO}{dt}$.

(c) Determine the direction of the arrow in each region.

Exercise 5-4

For the more general case

$$\frac{dV}{dt} = (\alpha - \gamma O)V, \qquad \alpha, \gamma > 0$$

$$\frac{dO}{dt} = (-\delta + \varepsilon V)O, \quad \delta, \varepsilon > 0,$$

(5-2)

(a) Determine where $\frac{dV}{dt} = 0$ and where $\frac{dO}{dt} = 0$.

(b) Determine the direction of the arrows in each of the four regions in the (V,O) plane according to the signs of $\frac{dV}{dt}$ and $\frac{dO}{dt}$.

(c) An equilibrium point of this process is a point at which $\frac{dV}{dt} = 0$ and $\frac{dO}{dt} = 0$. Find the nonzero equilibrium point for the model defined by Eqs. (5-2).

The arrows you have plotted in Exercise 4(b) should indicate that the process rotates about the equilibrium state. There are different possibilities that could occur. Two of these are shown in Figure 5-2.

In Figure 5-2(A), the process spirals in toward the equilibrium point, and in Figure 5-2(B) the process circulates about the equilibrium point.

Exercise 5-5

Describe how the populations of the owls and voles evolve in the two situations above.

In the problems so far, we have tried to understand the evolution of the owl and vole populations with respect to one another but have said nothing about how these populations evolve in time. In the next group of exercises, you are asked to determine the phase diagram of the owls versus the voles, given the graph of each species versus time.

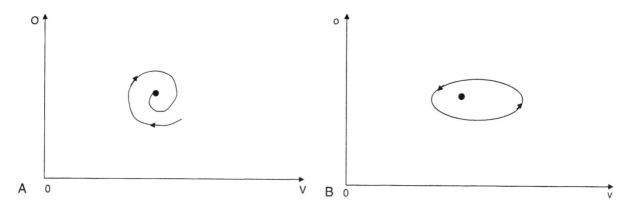

FIGURE 5-2.
Possible trajectories near an equilibrium point. Panel A: A trajectory that spirals into the equilibrium point; panel B: A trajectory that orbits the equilibrium point.

Exercise 5-6
....

The time trajectories in Figure 5-3(A) and Figure 5-3(B) describe the evolution of the owl and vole populations with time, respectively. In Panel C, sketch the corresponding phase-plane trajectory of voles against owls [i.e., sketch the trajectory of the process in the (V, O) plane].

Likewise, if we know the phase diagram and a point on the phase diagram that corresponds to a specific time (e.g., $t = 0$), we can graph the time trajectory for the population size of each species.

Exercise 5-7
....

Figure 5-4 gives a possible phase trajectory corresponding to a Lotka–Volterra model. Based on the information in Figure 5-4, determine the time trajectories for the sizes of the owl and vole populations. Sketch the graphs in Figures 5-5 and 5-6, respectively. Assume that at time $t = 0$, the owl and vole populations are at point A_1 [i.e., $O(0) = 30$ and $V(0) = 50$]. Assume also that it takes equal amounts of time for the trajectory to move between the points A_1, A_2, A_3, and A_4 marked on the graph, and that this time is equal to three hours.

Exercise 5-8
....

Enter the Lotka–Volterra model from Eq. (5-2) in *BEREKELEY MADONNA*, and run it with the following set of initial conditions and values of the parameters:

```
STARTTIME = 0

STOPTIME = 1000
```

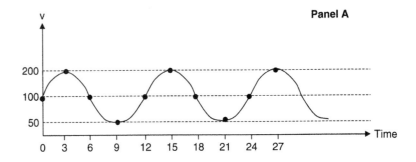

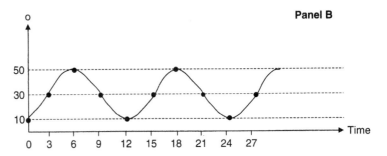

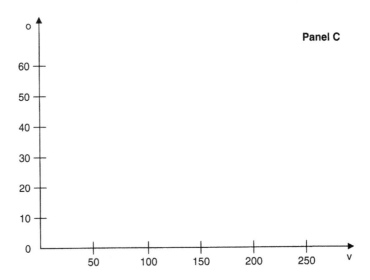

FIGURE 5-3.
Evolution of owl and vole populations. Panel A: Time trajectory of the owl population; panel B:
Time trajectory of the vole population; panel C: Template for sketching the trajectory of the voles
and owls in the (V,O) plane.

```
DT = 0.02

init O = 40

init V = 250

epsilon = 0.0002

delta = 0.02

alpha = 0.06

gamma = 0.001
```

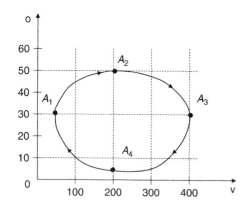

FIGURE 5-4.
A possible phase trajectory of a Lotka–Volterra model.

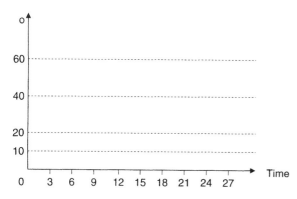

FIGURE 5-5.
Template for sketching the time trajectory of the owl population.

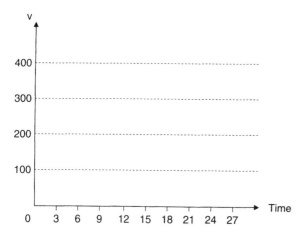

FIGURE 5-6.
Template for sketching the time trajectory of the vole population.

(a) Run the model to obtain the time trajectories of the numerical solutions for $O(t)$ and $V(t)$.

(b) Calculate the nonzero equilibrium state (V_0, O_0) of the Lotka–Volterra model for this choice of parameters. Do the time trajectories appear to approach equilibrium as $t \to \infty$? Why or why not?

(c) Display the phase-trajectories in the (V,O) plane by choosing Graph→Choose Variables from the main menu, selecting V as the x-axis variable, and removing V from the list of y-axis variables. Format the x-axis to display the range from 0 to 400 and the y-axis to display the range from 0 to 300.

(d) Recall that depressing the $\mathbf{I_c}$ button in *BERKELEY MADONNA* allows you to define new initial conditions $(V(0),O(0))$ in the phase plane by simply clicking at a point. The coordinates of this point in the (V,O) plane are then used as initial conditions $(V(0),O(0))$ for the model. Make a few clicks near the equilibrium point (V_0,O_0) determined in part (b). What do you expect to happen if you click exactly at the point (V_0,O_0)? Why?

(e) Test your answer from (d). Were you correct?

(f) Based on information from the phase trajectories, what is the type of the nonzero equilibrium point (V_0,O_0)? (circle one)

- Asymptotically stable

- Unstable

- Neutrally stable

- Repellor

(g) Save your work and close *BERKELEY MADONNA*.

PART II: OTHER PREDATOR–PREY MODELS

From our work with population models, we might recognize an immediate weakness in this model because it allows for exponential growth of voles in the absence of predators. Thus, we want a version of the Lotka–Volterra model that hypothesizes logistic growth for the voles with inherent per capita rate of growth α and carrying capacity of the environment K.

Exercise 5-9
....

Write the system of differential equations representing the mathematical model built upon the above hypothesis.

Exercise 5-10
....

Enter your model from Exercise 9 into *BERKELEY MADONNA* using the same values of the parameters and same initial conditions as in Exercise 8. Use $K = 400$ for the carrying capacity. In *BERKELEY MADONNA*, your model should look somewhat like

```
STARTTIME = 0

STOPTIME = 2000

DT = 0.02

d/dt(V) = alpha*(1 − V/K)*V −gamma*O*V

d/dt(O) = (−delta + epsilon*V)*O

init O = 40

init V = 250

epsilon = 0.0002

delta = 0.02

alpha = 0.06

gamma = 0.001

K = 400
```

Run the model.

   (a) Do the time trajectories appear to approach an equilibrium point (V_0, O_0) as $t \to \infty$? If so, estimate the point (V_0, O_0).

   (b) How is the long-term behavior of this model similar to or different from the long-term behavior of the Lotka–Volterra model?

   (c) Display the phase plot of the solution in the (V,O) plane. Describe the behavior of the phase trajectory near the equilibrium point.

   (d) Depress the I_c button and make several clicks to give different initial conditions for the model. Does the behavior of the phase trajectory near the equilibrium point (V_0, O_0) appear to change depending on the initial conditions?

   (e) Based on your observations in part (d), classify the equilibrium point (V_0, O_0) as stable or unstable.

Exercise 5-11

Define sliders on the model parameters, and observe the changes in the phase trajectories caused by changing the parameter values, one at a time.

   (a) Does the equilibrium point (V_0, O_0) change when the values of the model parameters are changed? If so, in what way?

(b) Does the nature of the equilibrium point (i.e., stable or unstable) change when the values of the model parameters are changed? Consider enough different combinations of parameter values to feel comfortable to build a hypothesis based on "experimental observations." A rigorous mathematical proof of the fact that when $\frac{\delta}{\varepsilon} < \frac{\alpha}{\beta}$, the state (V_0, O_0) is always stable regardless of the specific values of the parameters can be found in Chapter 2, Section VI. What happens when $\frac{\delta}{\varepsilon} \geq \frac{\alpha}{\beta}$?

Save your work and close *BERKELEY MADONNA*.

We next modify the model again in the following way:

1. We change the assumption regarding how the owls devour the voles so that instead of being proportional to VO, it will be proportional to $\frac{VO}{1+V}$. The idea here is that an owl can eat only so many voles before losing interest. The assumption for growth of the voles in the absence of predators is the same as before (i.e., they follow the logistic equation).

So we have

$$\frac{dV}{dt} = \alpha\left(1 - \frac{V}{K}\right)V - \gamma\frac{VO}{1+V} = \alpha V - \beta V^2 - \gamma\frac{VO}{1+V},$$

where $\beta = \alpha/K$.

2. We also change the assumptions on the rate of change of the owls so that

$$\frac{dO}{dt} = \delta\left(1 - \frac{\varepsilon O}{V}\right)O.$$

Note that if V were a constant, the above equation would be a logistic equation for the owls' population growth with a stable equilibrium equal to $\frac{V}{\varepsilon}$. This reflects the fact that the carrying capacity of the predator population is proportional to the number of prey. If $\delta > 0$ and $\frac{\varepsilon O}{V} > 1$, the owls will have a negative per capita growth rate and will be dying out.

We now follow Taubes: *Differential Equations Modeling in Biology*, and use his choice of model parameters for our next exercise. We take

$$\frac{dV}{dt} = \frac{2}{3}V - \frac{V^2}{6} - \frac{VO}{1+V}$$

$$\frac{dO}{dt} = \delta O\left(1 - \frac{O}{V}\right).$$

Exercise 5-12

Open *BERKELEY MADONNA*, enter the above model, and run it with the following values of the parameters:

STARTTIME = 0

STOPTIME = 1000

DT = 0.02

delta = 0.1

init O = 1

init V = 2

(a) Run the model, and format the y-axis to span the range from 0 to 3. Do the solution trajectories approach equilibrium? If so, estimate its value from the graph.

(b) Consider now the phase plot of the model solution in the (V,O) plane. Format the horizontal and vertical axes to span the ranges from 0 to 3. Using the I_c button, run the model with a number of different initial conditions. Make a conjecture regarding the stability of the equilibrium.

(c) Change the value of STOPTIME now to 10,000, because we shall need to take a look at the long-term behavior of the phase trajectories and thus need to have our model run for a longer time. Define a slider on δ using a small increment value (e.g., 0.001) for δ in the Sliders dialog box. Examine the behavior of the model phase trajectories near the equilibrium point for different values of δ. Is there a drastic change in this behavior for a specific value of δ (consider specifically values for δ from 0.07 to 0.1)? If so, describe the change, and estimate the critical value of δ.

(d) In Section VI of Chapter 2 of the textbook, we prove that for values of δ smaller than $1/12$, the equilibrium point under consideration becomes a repellor. Choosing different values for $0 < \delta < 1/12$ and different initial conditions, use *BERKELEY MADONNA* to illustrate this result.

REFERENCES

Taubes, C. H. (2001). *Modeling differential equations in biology.* Upper Saddle River, NJ: Prentice Hall.

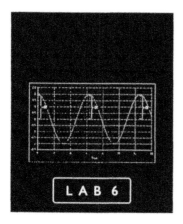

LAB 6

Selection in Genetics: The Effect of A Maladaptive or Lethal Gene

In this laboratory module, we examine the dynamics of a gene pool under the condition that one of the genotypes is less fit in comparison with the other genotypes. We use a discrete model to describe the change in genetic frequencies due to the selective disadvantage of the genotype with decreased fitness and study their long-term behavior. More specifically, in this laboratory exercise, you will:

- Explore the change in gene frequencies caused by selective disadvantage when a maladaptive or a lethal gene is present in the gene pool.

- Use difference equations to describe the change in gene frequencies from generation to generation.

- Determine the long-term behavior of gene frequencies.

- Document the elimination of the harmful allele due to natural selection.

BIOLOGICAL BACKGROUND

When Gregor Mendel performed his experiments in plant hybridization, he selected pea plants which differed at a single genetic locus and displayed alternate forms at that locus—tall or short plant height, yellow or green seed, inflated or constricted seed pod, and so on. He noted that when he crossed two true-breeding plants with alternate forms, such as tall and short, the offspring would represent only one of the two forms, in this case tall. However, when the offspring were allowed to reproduce, the new generation would also contain representatives with the missing form. Mendel realized that the causative agent for the missing form was actually present but masked by the presence of the other form. He postulated that the observed behavior was caused by the existence of "particulate factors," which we now call *genes* or, more properly, *alleles*. We call the masking allele *dominant* and symbolize it with a capital letter, such as **A**. We call the masked allele *recessive* and symbolize it with the corresponding lower case letter, **a**. Because each plant has two copies of each gene (one from each parent), we can describe the parents as **AA** × **aa** (which we call *homozygous*) and the offspring as all **Aa** (which we call *heterozygous*). Thus, we see that Mendel distinguished between the appearance (or *phenotype*) of the organism and its genetic complement (or *genotype*).

Early in the twentieth century, geneticists wondered why recessive alleles persisted from generation to generation. Why did the dominant allele not spread throughout the population and take over? The answer to this question was proposed independently by Godfrey H. Hardy and Wilhelm Weinberg, and is expressed in the principle that bears their names. It states that, under certain circumstances described by a series of specific assumptions, the genetic composition of the parental generation will be reflected unchanged in succeeding generations.

One of the most important assumptions of the Hardy–Weinberg principle is that no selection is occurring. When the fitness of all possible combinations of the recessive and dominant alleles at the gene locus is the same, the gene frequencies remain unaltered from generation to generation. The picture changes when some of the genotypes exhibit a selective disadvantage and are less fit to survive and reproduce. This is the case with certain genetic diseases.

What is selective disadvantage? Let us assume that the genotype **aa** confers a disease phenotype, such that only 80% of the **aa** individuals survive to the age of reproduction and reproduce. We assume that 100% of the **AA** and **Aa** individuals survive and reproduce. Therefore, the **aa** genotype confers a survival rate of 80% or $\alpha = 0.8$. This might be the case in a disease such as type I Gaucher disease caused by a recessive allele of the gene for the enzyme acid beta-glucosidase. The severity of the symptoms of type I Gaucher disease varies considerably, and improvements in current therapies could well result in survival and reproduction rates of 80% for affected individuals.

In general, the fraction α of the individuals with a given genotype that reaches reproductive age and thus takes part in the reproductive process producing the next generation of individuals is the genotype's *survival rate*. The difference $k = 1 - \alpha$ describes the disadvantage of the genotype in the selection process. The quantity k is the *selective disadvantage* of the specified genotype. In its extreme form, the selective disadvantage could be 100%. For example, Tay–Sachs disease is a devastating condition caused by a recessive allele that produces a defective version of the enzyme hexosaminidase A. Children born with two copies of the defective allele undergo a severe and progressive degeneration of nervous system function that, in general, results in death by age 3. Because no individuals with Tay–Sachs disease would live to reach reproductive age, the survival rate here will be $\alpha = 0$. Therefore, when a certain genotype exhibits a selective disadvantage, it appears that the Hardy–Weinberg equilibrium will be violated, and the frequencies p and q of the **A** and **a** alleles will no longer be maintained from one generation to the next. Further, if the **aa** genotype confers a selective disadvantage, then it is natural to expect that q—the frequency of the **a** allele in the gene pool—will be decreasing over time. If this is the case, then p—the frequency of the **A** allele—will be increasing.

To simplify matters for this exercise, we have been considering only the effects of the deleterious allele combination on survival. It is true, however, that a harmful allele combination might also have an impact on fertility, as is the case with diseases such as galactosemia or phenylketonuria. In either case, whether the mechanism involves reduced viability or reduced fertility, the result would be the same: The individual with the deleterious allele combination would be less likely to contribute those alleles to the next generation, and therefore the frequency of the deleterious allele in the gene pool should decline. As you consider the following exercises, however, keep in mind that a more inclusive name for α would be the "reproductive success" rate.

In the next section, we present a mathematical model that justifies the above heuristic claims, and we show how the values of p and q would change over time. Because the system is no longer in Hardy–Weinberg equilibrium, the frequencies p and q may change from generation to generation. We shall use p_n and q_n to denote the frequencies of the **A** and **a** alleles in the n-th generation.

MATHEMATICAL BACKGROUND

We would like to determine now in what way the genetic frequencies change from generation to generation and how the value of the survival rate α affects that change.

Suppose that the initial proportions of the alleles are:

$$p_0 = \text{proportion of } \mathbf{A} \text{ allele}$$
$$q_0 = 1 - p_0 = \text{proportion of } \mathbf{a} \text{ allele.}$$

Suppose also that the initial proportions of the phenotypes is as predicted by Hardy–Weinberg; that is

AA	**Aa**	**aa**
p_0^2	$2p_0q_0$	q_0^2

As we now suppose that the **aa** combination is harmful, only a fraction α (where $0 \leq \alpha < 1$) of the homozygous recessive genotype survives to reproduce. Thus, before reproduction, the genotype distribution in the population will change and will be

AA	**Aa**	**aa**
p_0^2	$2p_0q_0$	αq_0^2

It is clear that, the decreased amount of the **aa** genotype at the time of reproduction will cause a decrease in the **a** allele frequencies in the next generation. This could be heuristically justified by observing that if the total number of individuals in the original population was N (recall that we assumed N to be very large), there were Nq_0^2 homozygous recessive individuals to start with. Because only a fraction α of them survives to reproductive age, the number of those individuals has changed to $N\alpha q_0^2$ by the time of reproduction. Thus, the number of **A** alleles at that time will be $2p_0^2 N + 2p_0 q_0 N$ and the number of **a** alleles at that time will be $2\alpha q_0^2 N + 2p_0 q_0 N$. Denoting the proportions of the **A** and **a** alleles in the first generation by p_1 and q_1, we now calculate that

$$q_1 = \frac{2\alpha q_0^2 N + 2p_0 q_0 N}{2\alpha q_0^2 N + 2p_0 q_0 N + 2p_0^2 N + 2p_0 q_0 N} = \frac{\alpha q_0^2 + p_0 q_0}{\alpha q_0^2 + 2p_0 q_0 + p_0^2} = q_0 \left(\frac{\alpha q_0 + p_0}{\alpha q_0^2 + 2p_0 q_0 + p_0^2} \right).$$

The proportion p_1 could be calculated similarly, but because we assumed that **A** and **a** are the only alleles present in the population, it is easier to use that $p_1 = 1 - q_1$.

Similarly, if p_n and q_n denote the proportions of the **A** and **a** alleles in the n-th generation, we obtain that because of weaker fitness of the **aa** genotype these frequencies will change in the $(n + 1)$-st generation to

$$q_{n+1} = q_n \frac{\alpha q_n + p_n}{\alpha q_n^2 + 2p_n q_n + p_n^2}$$

and

$$p_{n+1} = 1 - q_{n+1}. \tag{6-1}$$

Notice how different this situation is in comparison with the Hardy–Weinberg case. The allelic frequencies of **A** and **a** now change from generation to generation, and when we know the frequencies for any given generation, Eq. (6-1) allows us to compute their values for the following generation. Expressions such as Eq. (6-1) are called *recursive formulae*. It is easy to notice a disadvantage of such formulae—in order to calculate the values of p_n and q_n for, say $n = 100$, we need to calculate the values of p_1 and q_1 from p_0 and q_0, then the values of p_2 and q_2 from those of p_1 and q_1 obtained on the previous step, and so on, all the way to p_{100} and q_{100}.

Luckily, such formulae are very easy to program for a computer, and thus the calculations present no problem.

Notice that because

$$\alpha q_n^2 + 2p_n q_n + p_n^2 = (\alpha q_n + p_n)q_n + p_n(q_n + p_n) = (\alpha q_n + p_n)q_n + p_n,$$

Eq. (6-1) could be rewritten as

$$q_{n+1} = q_n \frac{\alpha q_n + p_n}{(\alpha q_n + p_n)q_n + p_n}, p_{n+1} = 1 - q_{n+1}. \qquad (6\text{-}2)$$

Now, because $\alpha q_n < \alpha q_n + p_n$, we see that $\dfrac{\alpha q_n + p_n}{(\alpha q_n + p_n)q_n + p_n} < 1$, and, together with Eq. (6-2), this proves that $q_{n+1} < q_n$. Thus, the allelic frequency of the harmful allele **a** does indeed decrease from generation to generation.

The dependence described by Eq. (6-2) can easily be formulated in terms of the selective disadvantage k of the homozygous recessive genotype. Recalling that $k = 1 - \alpha$, Eq. (6-2) becomes

$$q_{n+1} = q_n \frac{(1 - k)q_n + p_n}{((1 - k)q_n + p_n)q_n + p_n}, p_{n+1} = 1 - q_{n+1}. \qquad (6\text{-}3)$$

In Chapter 3 of the textbook we give a detailed proof that, in the long run, the frequencies q_n approach zero. Thus, we have established the following fact: *Under the assumption that the recessive homozygous genotype* **aa** *is less fit than the homozygous dominant* **AA** *and the heterozygous* **Aa** *genotypes, the frequency of the harmful allele* **a** *will diminish from generation to generation, and the allele* **a** *will be eventually eliminated from the gene pool.*

Knowing that the frequency of the harmful allele will decrease to zero over time, an important question now is to ask how many generations it will take for the maladaptive allele to reach virtually negligible frequency levels in the gene pool. The exercises in this laboratory module provide a glimpse into this important issue.

SOFTWARE BACKGROUND

In this laboratory module, we shall work with difference equations and recursive formulas. Having used *BERKELEY MADONNA* for all previous laboratory projects, we shall continue to do so here. However, all calculations could also be carried out with equal ease in *MS Excel* or any other spreadsheet application.

More specifically, in this module we use *BERKELEY MADONNA* to generate the sequence $q_0, q_1, q_2, q_3, \ldots$ defined via the recursive Eqs. (6-2) and (6-3), where q_0 is the initial frequency of the harmful allele. As Eqs. (6-2) and (6-3) define a discrete model, the model should be entered into *BERKELEY MADONNA* in the form of difference equations. For example, if it is known that $q_0 = 0.3$, and the survival rate is $\alpha = 0.7$, Eq. (6-2) can be entered into *BERKELEY MADONNA* as:

```
STARTTIME = 0

STOPTIME = 10
```

```
DT = 1

init q = 0.3

next q = q*(alpha*q + p)/((alpha*q + p)*q + p)          (6-4)

p = 1 - q

alpha = 0.7
```

The value of STOPTIME in this case corresponds to the number of generations over which we wish to examine the changes in allele frequencies. The code presented above will compute the values of q_n and p_n for $n = 1,2,3,\ldots,10$. The computed values will be displayed in a form similar to that shown in Table 6-1.

Recall that to be able to see the table of values in *BERKELEY MADONNA*, you shall need to Run the model and then depress the Table button. If you need a review on how to do this, revisit Exercise 1, Lab 1.

Exercise 6-1

Assume that, initially, the frequency of the harmful allele **a** is $q_0 = 0.5$. (How might a harmful allele reach such a high frequency in the population? It might do so in a situation such as occurs with sickle-cell anemia, where individuals homozygous for the normal allele are subject to malaria, and heterozygous individuals are protected against malaria. If a small population of heterozygotes were to migrate to a malaria-free area, we might encounter a situation such as this.) If the survival rate of the homozygous recessive genotype **aa** is $\alpha = 0.65$, use *BERKELEY MADONNA* and Eq. (6-2) above to calculate after how many generations the frequency of the **a** allele will fall below

Generation	q_n	p_n
0	0.300000	0.700000
1	0.280576	0.719424
2	0.263174	0.736826
3	0.247539	0.752461
4	0.233448	0.766552
5	0.220707	0.779293
6	0.209150	0.790850
7	0.198634	0.801366
8	0.189034	0.810966
9	0.180247	0.819753
10	0.172178	0.827822

TABLE 6-1.
The Values Obtained from the Recursive Formula in Eq. (6-4) and $q_0 = 0.3$ for the First Ten Generations

(a) 0.25?

(b) 0.05?

(c) 0.005?

Exercise 6-2

The frequency of the Tay–Sachs allele varies significantly among human population groups. It is very high among Ashkenazi Jewish families and relatively low among non-Jews and among Jews who are not of Askenazi extraction. The frequency of this allele in the Jewish population of New York City is believed to be about 0.016. As noted in the introduction, it is highly unlikely for individuals with Tay–Sachs disease to live to reach reproductive age; thus, the survival rate α can be assumed to be $\alpha = 0$.

(a) Find the number of generations it would take for this frequency to fall to 0.010. Repeat for a frequency of 0.001.

(b) If one assumes that a generation is 25 years, how many years do these two values of n represent?

(c) Just to put things into perspective, find out how long the species *Homo sapiens sapiens* (us) has existed and compare with the results from part (b).

Exercise 6-3

As is illustrated by the Tay–Sachs example in Exercise 6-2, the frequency of harmful alleles is usually quite small, even in diseases which occur quite frequently. For example, consider cystic fibrosis (CF), which is extremely common among white Americans. The frequency of the CF allele among the Old Order Amish of Holmes County, Ohio, was estimated to be 0.042. This results in an incidence of 1 in 569 live births, which is much higher than the rate in the rest of the white population of Ohio. The higher rate of the harmful allele reflects the migration of a relatively small population of Amish individuals to Holmes County, and the tendency of Amish people to marry within their group rather than with the members of the population at large.

If the homozygous recessive gene **aa** carries a selective disadvantage $k = 0.45$ (which, in the case of CF, would be caused by both early death and reduced fertility), how many generations will be needed for the frequency of the **a** allele to fall from its current value $q_0 = 0.042$ to a value below

(a) 0.01?

(b) 0.005?

(c) 0.001?

Exercise 6-4
....

Repeat Exercise 6-3 for the case in which the selective disadvantage k is only 15%. What changes do you observe when the value of k changes? Suggest some biological explanations that could account for the decreased value of k (**Hint:** Consider the general attitude of the Amish culture toward modern American society, then consider a population with a different attitude). Could you formulate a general result linking the dependence between the rate of change in the allele frequency q and the magnitude of the selective disadvantage k carried by the **aa** genotype?

Exercise 6-5
....

Thus far, we have been modeling diseases caused by autosomal recessive alleles. Now, let us look at an autosomal dominant disease, such as Long QT syndrome or Marfan syndrome. Long QT syndrome is an inherited tendency to cardiac arrhythmias which, when triggered by stressors such as exercise, can cause death. Marfan syndrome is a defect in connective tissue, which results in cardiovascular abnormalities as well as skeletal deformities and imperfect eyesight. In either of these diseases, the homozygous recessive genotype would be completely normal—conferring, in fact, a greater chance of survival than the heterozygous and the dominant homozygous genotypes.

Suppose that homozygotes with the recessive allele have a 30% greater chance of survival than either **AA** or **Aa**.

(a) How would you expect the frequency of the allele **a** to change over time?

(b) If the initial frequency of **a** is $q_0 = 0.95$, find the number of generations needed before the gene frequency of **a** reaches 99%.

Hint: View the selective advantage of 0.3 as a negative selective disadvantage.

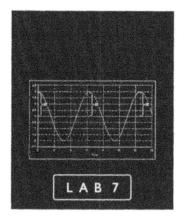

LAB 7

Quantitative Genetics and Statistics

In this laboratory module, we examine quantitative differences between samples and the problem of inheritance of traits. In this laboratory module, you will:

- Formulate hypotheses and understand how to apply statistical methods to perform hypothesis testing.

- Practice hypothesis testing in the context of comparing mean values between samples from different populations.

- Use *MINITAB* to perform some standard statistical analyses.

BIOLOGICAL BACKGROUND

A fundamental question, both philosophically and biologically, is what causes differences within a species. The issue is complicated, but it is common to classify the causes as genetic, environmental, or developmental "noise." In general, noise may be considered to be a random occurrence, unrelated to the factor(s) under consideration. It may be illustrated by the fact that identical twins, when they emerge from the womb, will have some differences. When considering genetic, environmental, or developmental factors, one would like to be able to attribute certain amounts of influence to each of these factors, but this is usually not possible on an individual basis. Complicating the problem further, there will usually be some crossover between factors (called *covariance*). As an example, a child whose parents are musicians may be genetically predisposed toward music, if such genes exist, but would also have the benefit of being reared in an environment that is musically nurturing. Similar confounds occur in analyzing intelligence levels. More information about these and other questions of interest to quantitative genetics can be found in Falconer (1989).

In this laboratory project, we examine some statistical methods that enable us to make an unbiased judgment on a hypothesis based on collected data. In particular, we shall focus on the following related topics:

1. How do we decide whether any observed differences are statistically significant?

2. Is the contribution of an underlying genetic or other factor significant, relative to the contribution of environmental noise?

MATHEMATICAL BACKGROUND

Probability. In this laboratory exercise, we shall work with *continuous* random variables. In the case of continuous random variables, the *probability density functions* are used to compute probabilities. A function $f(x)$ is a probability distribution density function for a random variable ξ provided that:

1. $f(x) \geq 0$

2. $\int_{-\infty}^{\infty} f(x)dx = 1$

3. For any numbers $a < b$, the probability that ξ is between a and b is calculated as $P(a < \xi < b) = \int_{a}^{b} f(x)dx.$

The most common and important probability density function is that of the so-called *normal* or *Gaussian distribution*. A very special case is the *standard normal density function* defined by the expression

$$f(x) = \frac{1}{\sqrt{2\pi}} e^{-x^2}. \tag{7-1}$$

The graph of this function is depicted in Figure 7-1(A).

The *mean* and the *variance* of a random variable are two numerical characteristics associated with its distribution. The mean could be thought of as the average we would expect after doing many trials. The variance measures the spread of the data around the mean value. The mean is typically represented by μ and the variance by σ^2. The *standard deviation*, defined as the square root of the variance, is denoted by σ.

In Chapter 3 of the text, we saw how a normal distribution can be obtained from a binomial distribution as the number of observations increases. We introduced the familiar bell-shaped curve of the Gaussian distribution that achieved its maximal value at the mean value μ of the distribution. This was affirmed by the Central Limit Theorem, guaranteeing that for sufficiently large values of m, the histogram of the binomial distribution is well approximated by a bell curve that has the same mean and variance as the binomial distribution.

The question of why this happens is quite general in nature. Although by no means elementary, proofs of the Central Limit Theorem, together with the specific conditions under which it holds, can be found in almost any standard text in probability. Because the arguments are technical in nature and require a considerable amount of probability background, we shall not present them here. Instead, the following heuristic explanation could be used to outline the main principle: A normal distribution occurs when multiple independent random choices are made, each of them attempting to achieve a certain fixed average value, but each one vulnerable to errors that are symmetrical in both directions around the mean.

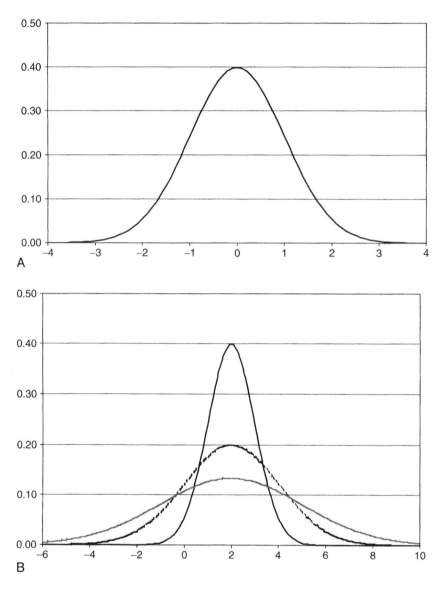

FIGURE 7-1.
Comparison of normal density functions with differing parameters. Panel A: Standard normal density (i.e., normal density with mean $\mu = 0$ and standard deviation $\sigma = 1$); panel B: normal densities with $\mu = 2$ and $\sigma_1 = 3$ (gray line), $\sigma_2 = 2$ (black dotted line), and $\sigma_3 = 1$ (solid black like).

Several examples would clarify this point. Thousands of bats exit their cave about 2 hours after sunset. Because inside the cave there is no indication of exactly when the sunset occurs, each bat relies on its biological clock to estimate the exit time. It is a fact that some biological clocks run faster and some lag behind, with larger errors less likely to occur than smaller ones. Experiments have shown that the number of bats exiting a cave per minute follows quite precisely a bell-shaped curve with a mean of about 2 hours after sunset.

Similarly, if we place microscopic particles in a glass of water, they get hit by the water molecules and travel various distances depending on the force of the impact. The average travel distance depends on the temperature (energy) of the water, and the distribution of the distances around this average is approximately normal. In other words, each bat's time estimate of 2 hours after sunset or each particle's travel distance is one outcome of an experiment described by a random variable with normal density function. The general analytical expression of the normal (also called Gaussian) density is given by the function

$$f(x) = \frac{1}{\sqrt{2\pi} \cdot \sigma} e^{\left(\frac{x-\mu}{\sigma}\right)^2}, \tag{7-2}$$

where x can take any value in the interval $(-\infty, \infty)$. When we say that a random variable ξ has a normal distribution with parameters μ and σ, this means that the random quantity represented by ξ has a density function such as in Eq. (7-2). The values $\mu > 0$ and $\sigma > 0$ uniquely define the position and shape of the curve and correspond to the mean and the standard deviation of the normal distribution. The mean value μ determines the position of the maximum for the bell-shaped graph of the normal density function, while the standard deviation σ determines how sharp the peak is near the maximum. Figure 7-1(B) illustrates how the density graphs for the function in Eq. (7-2) change with the change of the parameters μ and σ. Notice that when $\mu = 0$ and $\sigma = 1$, the density function from Eq. (7-2) takes the form presented by Eq. (7-1). That is, for $\mu = 0$ and $\sigma = 1$, we obtain the *standard normal distribution*.

In addition to the normal distribution, the following distributions will also be used in this lab.

(1) χ^2 *(Chi-square) Distribution.* This type of distribution arises when we consider squares of random variables with standard normal distribution. More specifically, if ξ is a random variable with a standard normal distribution, it may sometimes be necessary to consider its square, that is, $\eta = \xi^2$. Because ξ is a random variable, so is η, but the density function of η cannot be normal because the new random variable η takes only positive values. We say that $\eta = \xi^2$ has a χ^2 *distribution with one degree of freedom.* If we consider several independent random variables, $\xi_1, \xi_2, \ldots, \xi_N$, all of which have standard normal distributions, then we say that the sum of the squares of these random variables, namely, $\eta = \xi_2^1 + \xi_2^2 + \ldots + \xi_N^2$, has a χ^2 *distribution with N degrees of freedom.* Figure 7-2(A) presents the density functions of χ^2 distributions with $N = 7$ and $N = 21$ degrees of freedom, respectively.

(2) t *(Student's) Distribution.* This distribution arises when we need to consider specific ratios of random variables. In particular, if ξ is a random variable with a standard normal distribution and η is another random variable with a χ^2 distribution with N degrees of freedom,

then their ratio $\zeta = \xi / \sqrt{\eta}$ will be a new random variable. This new random variable is said to have a *t-distribution* (or *Student's distribution*) *with N degrees of freedom*. The graphs of the probability density for a *t*-distribution with $N = 2$ and $N = 23$ degrees of freedom are shown in Figure 7-2(B).

(3) *F (Fisher) Distribution*. This distribution also arises when ratios of random variables are considered, but this time it is the ratio of two random variables with χ^2 distributions. More specifically, if ξ is a random variable with a χ^2 distribution with M degrees of freedom and η is another random variable with a χ^2 distribution with N degrees of freedom, then their ratio $\zeta = \xi / \eta$ will be a new random variable said to have an *F-distribution* (or *Fisher distribution*) *with M, N degrees of freedom*. The graphs of the probability densities for two F-distributions with different degrees of freedom are shown in Figure 7-2(C).

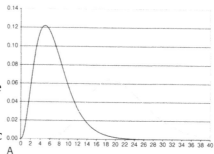

Statistics. We now explain how these probability definitions relate to common problems of statistical testing. We shall do this at a general heuristic level, without getting into rigorous mathematical proofs that can be found in standard statistical textbooks (see, for example, Devore and Peck 2005).

The mathematical formulation of statistical problems uses some specific language that we outline next. It is common for a statistical problem to begin like this: Let $x_1, x_2, \ldots, x_N$ be N independent observations of a normally distributed random variable ξ with unknown parameters μ and σ. This means that we have collected some data by measuring a random quantity known to have a normal distribution, and the values from the measurements have been denoted by $x_1, x_2, \ldots, x_N$. These values are the *data points* that form our sample from the random variable ξ.

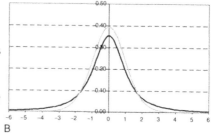

The mean of a random variable could be thought of as the average after many trials. Thus, it is natural to expect that the average value of the data points, denoted by $\bar{x}$,

$$\bar{x} = \frac{x_1 + x_2 + \ldots + x_N}{N} = \frac{\sum\limits_{i=1}^{N} x_i}{N} = \frac{1}{N} \sum\limits_{i=1}^{N} x_i, \tag{7-3}$$

would be a good estimate of the mean value parameter μ of the normal distribution.[1] This is the reason why the average value calculated in Eq. (7-3) is sometimes called the *empirical mean* or the *sample mean* of the random variable ξ. It can be shown that this is the best (in terms of statistical criteria) estimate, also called the *maximum likelihood* estimate for the mean μ. In these terms, the test average that you compute in a

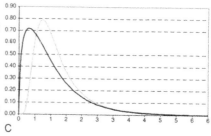

FIGURE 7-2.
χ^2, *t*, and *F* distributions with varying parameters. Panel A: χ^2 distribution with 7 (black) and 21 (gray) degrees of freedom; panel B: *t*-distribution probability density with 2 (black) and 23 (gray) degrees of freedom; panel C: *F*-distribution with 3, 40 (black) and 23, 8 (gray) degrees of freedom.

---

1. As is customary in mathematics, we have used $\sum_{i=1}^{N} x_i$ to denote the sum $x_1 + x_2 + \ldots + x_N$.

class is a maximum likelihood estimate of your grade. Similarly, it can be shown that a maximum likelihood estimate s^2 of the variance σ^2 of a normal distribution is given by the formula

$$s^2 = s^2(N) = \frac{(x_1 - \bar{x})^2 + (x_2 - \bar{x})^2 + \ldots + (x_N - \bar{x})^2}{N - 1}$$

$$= \frac{1}{N - 1} \sum_{i=1}^{N} (x_i - \bar{x})^2. \tag{7-4}$$

The value s^2 is sometimes called the *empirical variance* or *sample variance*. The square root, s, of this value is called the *empirical standard deviation* of the random variable ξ.

If we need to measure two different random variables ξ and η that have normal distributions, we record the data points for ξ and η as samples A and B. The sample A contains the data points measured for ξ and sample B contains the points measure for η. To distinguish between the mathematical expressions using data from sample A from those using data from sample B, we use the name of the sample as part of the notation. For example, we use $\bar{x}_A$ to denote the average of the data points from A, and $s_B^2 = s_B^2(N)$ to denote the maximum likelihood estimate of the variance calculated by the formula in Eq. (7-4) with the data from sample B. In the latter case, the value of N will correspond to the number of data points in the sample B.

Recall that the sum of two independent, normally distributed variables also has a normal distribution with mean value equal to the sum of the means and variance equal to the sum of variances. It follows that the empirical mean $\bar{x}$ and the difference $(x_i - \bar{x})$ will also have normal distributions. The same holds for the difference $\bar{x}_A - \bar{x}_B$, when we consider two samples. Particularly, $\bar{x}$ will have a mean equal to μ and a variance equal to σ^2/N.[2] Further, because $(x_i - \bar{x})$ has a normal distribution, the estimate of the variance s^2 (which is a sum of the squares of N such quantities) will have an approximately χ^2 distribution with N degrees of freedom.

When these considerations are paired with the definitions for the χ^2 distribution, t-distribution, and F-distribution we gave above, it follows that the following broad principles hold:

1. Any statistical test (such as the Z-test, as we'll see later) that uses the difference $\bar{x}_A - \bar{x}_B$ between the empirical means of two samples A and B as their test statistic would require the use of a normal distribution;

---

2. For the two different samples, A and B, the difference $\bar{x}_A - \bar{x}_B$ will be normally distributed for the same reason.

2. Any statistical test (such as the *t*-test, as we'll see later) that uses a variant of the ratio $\bar{x}/s$ (empirical mean/empirical standard deviation) as their test statistic, would require the use of a *t*-distribution (as it is approximately normal/$\sqrt{\chi^2}$); and

3. Any statistical test (such as the *F*-test, as we'll see below) that uses the ratio $s_A^2(M)/s_B^2(N)$(empirical variance of one sample with *M* readings/empirical variance of another sample with *N* readings) as their test statistic, would require the use of a *F*-distribution (as it is approximately χ^2/χ^2) with *M, N* degrees of freedom.

Testing a Hypothesis. The idea of hypothesis testing is that we make a claim and then evaluate the likelihood of this claim based on data. The claim is called the *null hypothesis*. The negation of the null hypothesis is called the *alternative hypothesis*. For example, suppose the null hypothesis is that the mean value μ of a sample has the probability distribution presented in Figure 7-3(A). Then we take a sample of size 20 and find the sample mean $\bar{x}$ is 0.7. In Figure 7-3(B), we locate the particular value for our sample and shade the area under the probability density curve to the right of this value. The area of the shaded region represents the probability the sample mean we took would have been 0.7 or larger, if the null hypothesis is correct. This probability is the *p*-value corresponding to the empirical result of 0.7. A small *p*-value would indicate the observed average is unlikely to have occurred among all samples of size 20 if the null hypothesis were true, and therefore provides evidence for rejecting the null hypothesis.

Assume now that we want to compare the mean values of two populations, A and B. We may think of this in the context of two types of corn, regular and genetically engineered, and want to know whether the genetically engineered corn variety B produces a higher average yield than the original variety A. The yields from A and B can be considered random quantities, and we denote them by ξ and η, respectively. We denote the mean values of ξ and η by μ_A and μ_B,

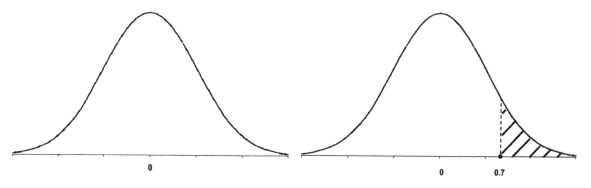

FIGURE 7-3.
Standard normal distribution. Panel A: Probability distribution for the means of samples of size 20, if the null-hypothesis is correct; panel B: The shaded region represents the probability that, if the null-hypothesis is correct, the sample mean for a sample of size 20 is 0.7 or larger.

respectively. In this case, rejecting a null hypothesis that $\mu_A \geq \mu_B$ will provide supporting evidence for the alternative $\mu_A < \mu_B$, which would imply evidence in favor of higher yield by the genetically engineered variety. It is common to denote the null hypothesis by H_0 and the alternative hypothesis by H. In general, hypothesis tests are classified as *one-tailed* or *two -tailed*, or, alternatively, as *one-sided* and *two-sided*. A one-tailed test is a test in which the hypothesis is directional (i.e., uses either a "<" or a ">." The corn problem above is an example of a one-tailed test. In a two-tailed test, the hypothesis does not specify a direction and will use the "$\neq$" symbol. An example of a two-tailed test with the same data would be "Is there a difference between the yield of plants A and B?" In this case, the null hypothesis would be H_0: $\mu_A = \mu_B$, and H: $\mu_A \neq \mu_B$ would represent the two-tailed alternative.

It is important to note that the decision one makes in hypothesis testing is to "reject the null hypothesis" or to "not reject the null hypothesis." Note that one does not use the language "accept the alternative hypothesis" even though that would seem appropriate. The analogy often used to explain this philosophy is a criminal case in the United States court system where the null hypothesis is "innocence." A jury will vote to convict only if there is compelling evidence of guilt. A vote to acquit does not imply that innocence was proven.

In hypothesis testing, there are two types of error that one can make. A *type I error* is rejecting a true null hypothesis, and a *type II error* is failing to reject a false null hypothesis. A type I error is what we really want to avoid. In the court analogy, it would be like convicting an innocent person (a type II error would be to acquit a guilty person). In the example above, the probability for type I error was depicted in Figure 7-3(B) and represented the probability for rejecting a true null hypothesis. It is generally accepted that this p-value should be small, typically less than 0.05, in order for the null hypothesis to be rejected. The number $1 - p$ is the *confidence level* of the test. The generally accepted confidence level of 0.95 is derived from the following optimization: It is intuitively clear that the lower we set the threshold p for type I error, the more confident we are that we can correctly reject the null hypothesis in favor of the alternative hypothesis. This, however, leads to the danger of type II error. Thus, the confidence level has to be optimized to balance the sample size and its associated type I and type II errors.

We now turn back to our corn example. Suppose that we are testing H_0: $\mu_A \geq \mu_B$ versus the alternative H: $\mu_A < \mu_B$ or, equivalently, H_0: $\mu_A - \mu_B \geq 0$ versus the alternative H: $\mu_A - \mu_B < 0$. Suppose also that when testing our hypothesis, we want to limit the magnitude of type I error that we may be making to 0.05. From the data gathered for the experiment, we compute $\bar{x}_B - \bar{x}_A$, which we denote by α. According to the procedure outlined above, we next need to evaluate the probability of how likely it is that the sampling distribution for $\bar{x}_B - \bar{x}_A$, determined based on assuming the correctness of the null hypothesis, would produce that value. Assume now that the sampling distribution for $\bar{x}_B - \bar{x}_A$ is represented by the density function depicted in Figure 7-4(A). Because we are considering a one-sided hypothesis, we find a value V_0 such that the area under the curve to the right of V_0 is exactly 0.05 [see Figure 7-4(B)]. Next, we plot the value $\alpha = \bar{x}_B - \bar{x}_A$ on the horizontal axis. If the value of α falls to the right of V_0, this would mean that if we reject the null hypothesis, the chances that we are wrong (that is, the chances that we shall reject the null hypothesis when it is, in fact, true) would be smaller than 5%.

If we have a two-sided alternative hypothesis H: $\mu_B \neq \mu_A$, we need to find two numbers, V_1 and V_2 [see Figure 7-4(C)], such that the area to the left of each number is 0.025 and thus the total area under the curve outside of the interval $[V_1, V_2]$ is 0.05. If the value $\alpha = \bar{x}_B - \bar{x}_A$ falls to the right of V_2 or to the left of V_1, then the chance that we shall wrongly reject the null hypothesis will be less than 5%.

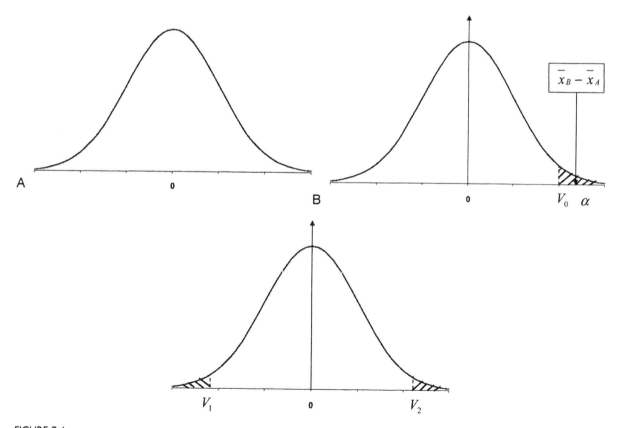

FIGURE 7-4.
Genesis of one-tailed and two-tailed p-values. Panel A: Sampling distribution of $\bar{x}_B - \bar{x}_A$. Panel B shows the one-tailed and panel C shows the two-tailed p-values.

We shall point out that the *p*-value produced by statistical software such as *MINITAB* and *SPSS* is equal (in the case of one-sided hypotheses) to the area under the curve to the right of the value of our statistic $\alpha = \bar{x}_B - \bar{x}_A$. Therefore, if this *p*-value is less than 0.05, this means that $\alpha = \bar{x}_B - \bar{x}_A$ is to the right of the value V_0 and therefore H_0 can be rejected with a chance of type I error less than 5%. In the case of a two-sided hypothesis, the *p*-value represents the combined area under the curve outside of the interval $(-|\bar{x}_B - \bar{x}_A|, |\bar{x}_B - \bar{x}_A|)$. The important thing to remember is that, in all cases, the *p*-value is exactly the probability for type I error.

One question that we have not addressed so far is how to determine the actual type of the sampling distributions under the assumption that the null hypothesis is true. This choice is based on underlying assumptions for the populations as well as on the type of parameters and claims referenced by the null hypothesis. As we shall see, the probability distributions that we introduced above play a fundamental role in the process. The following cases will be quite common:

- *Case I.* The null hypothesis deals with comparing mean values; and

- *Case II.* The null hypothesis deals with comparing variances.

As we already saw, the first case allows for answering questions similar to that of assessing the superiority in terms of yield of one corn type over another. This is most commonly done by using a Z-test or Student's *t*-test. Assume, as above, that we want to evaluate the claim H_0: $\mu_A \geq \mu_B$ (or, equivalently, $\mu_A - \mu_B \geq 0$) versus the alternative H: $\mu_A < \mu_B$ (or, equivalently, $\mu_A - \mu_B < 0$). In this case, the sampling distribution of $\bar{x}_B - \bar{x}_A$ will be approximately normal, and we would use either a Z-test or a *t*-test. We use a Z-test when the variances of each group are known. We use a *t*-test when the variance of either group is unknown. The latter requires the involvement of an estimate of the sample variance that leads to significant complications and the need to use a *t*-distribution as a sampling distribution of the test.

The second case will allow us to address the question of heritability, and we next illustrate how to do this. We need to decide whether the contribution of an underlying genetic or other factor is significant, relative to the contribution of environmental effects. In order to answer this question, we have to compare the variance explained by the genotype to the entire variance of the observed phenotype and to decide whether the genotype explains a significant portion of the entire variance. This brings us to one of the most important markers evaluated in genetic studies—the metric called *heritability*. Heritability is defined as the ratio of additive genetic variance (V_A) to the entire variance observed in the phenotype (V_P) and by tradition is denoted by $h^2 = V_A/V_P$. Numerous studies have been designed to evaluate the heritability of various traits. For example, the stature of a person is a trait with relatively high heritability, $h^2 = 0.65$, while insulin resistance, a major factor in the development of Type 2 diabetes, has heritability $h^2 = 0.31$ (see Bergman et al. 2003).

In general, the methods used to estimate heritability include parent–offspring regressions and/or analysis of variance (ANOVA) that compare siblings to half-siblings, or identical versus nonidentical twins. For the purposes of this project, the important property of heritability is that it is defined as the ratio of two variances. Therefore, given the probability considerations that we gave in the beginning, we would expect that in statistical problems heritability would have F-distribution.

Assuming that we have data on a specific trait collected for a sample of children together with data for the same trait from their parents, we can evaluate heritability via linear regression. Because we want to investigate the dependence of the child's characteristic on his or her parents' values, we choose the parental data to be the independent variable (denoted by X) and the child's data to be the dependent variable (denoted by Y).

Considering the (X,Y) scatter plot is always worthwhile, because it may be suggestive of a general linear relationship $Y = aX + b$ between the variables X and Y. Figure 7-5 illustrates this point and also includes the line that best fits the data set. This line is called the *least-squares regression line* for the data. A criterion for best fit and how the coefficients a and b for this line can be determined from the data can be found in Chapter 8 of the textbook. Denote the vertical distances of the data points from the line $Y = aX + b$ by $r_1, r_2, \ldots, r_n$. These numbers, calculated as $r_i = |Y_i - (aX_i + b)|$, $i = 1, 2, \ldots, n$, give the variation in the Y variable from the straight line relationship (see Figure 7-5) and are called *residuals*. The *sum of squared residuals (SSR)*, defined as

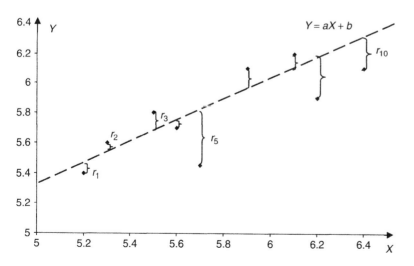

FIGURE 7-5.
Scatter plot of the (X,Y) data with a plot of the least-squares regression line and residuals.

$$SSR = r_1 + r_2 + \ldots + r_n = \sum_{i=1}^{n} r_i^2, \qquad (7\text{-}4)$$

is the most frequently used measure to express the combined variance of the data from the regression line.

The regression line in the figure represents the mathematical model that explains the variance in the data because of genetic factors. The value of SSR, on the other hand, represents the variance caused by other factors.

A second sum of squares, often called the *total sum of squares (TSS)* can be used to assess the total variation among the observed Y values. It is calculated as the sum of the squared residuals around the mean $\overline{Y}$ of the Y values (see Figure 7-6):

$$TSS = (Y_1 - \overline{Y})^2 + (Y_2 - \overline{Y})^2 + \ldots + (Y_n - \overline{Y})^2 = \sum_{i=1}^{n} (Y_i - \overline{Y})^2.$$

It can be shown (and it is somewhat obvious from the graphs) that for any set of points, $SSR \leq TSS$, and the equality is only possible when the least-squares regression line is the horizontal line $Y = \overline{Y}$. The difference $TSS - SSR$ gives the variance explained by the model, which, in this case, is the regression line. The coefficients of the least-squares regression line, together with the quantities SSR and TSS, can be obtained as part of the regression output from all standard statistical software packages.

The heritability ratio $h^2 = V_A/V_P$ (which represents the genetic variance divided by the entire variance of the phenotype) translates, in terms of our notation, to

$$h^2 = V_A/V_P = (TSS - SSR)/TSS.$$

The next question to consider is whether this result is statistically significant. More specifically, we want to see whether the portion of the variance explained by the regression (genotype) is statistically significant. We need to formulate a null hypothesis H_0 about the regression line and define a statistic that would allow us to decide whether we can reject it. The hypothesis that we would like to reject is:

H_0: *There is no genotypic (inherited) component in the child's stature caused by the parental stature.*

Mathematically, this would correspond to a horizontal regression line. Therefore, the hypothesis H_0 can be stated in mathematical terms as:

H_0: *The slope of the regression line is equal to zero.*

If H_0 were true, the value of SSR would be equal to the value of TSS and the variance in the data explained by the regression would be zero (see Figure 7-6).

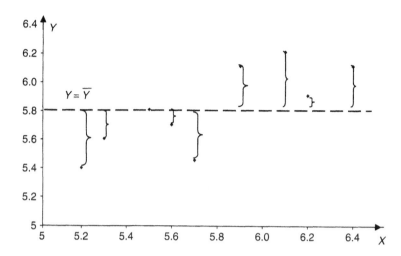

FIGURE 7-6.
Residuals of the Y values around the mean $\bar{Y}$ of all Y values.

Furthermore, recall that in that case, the mean square error of the residuals would have a χ^2 distribution. Therefore, the quotient (regression mean square error)/(residual mean square error), which is a quotient of two χ^2 distributions, would have an F-distribution.

The value of this quotient is calculated next to give the F-value for the test. Following the procedure illustrated with Figure 7-4(C), we ask what is the probability of obtaining such a value if H_0 were true. The answer is found using software that computes the F-distribution, or from F-distribution tables, taking into account the degrees of freedom we determined. The p-value, corresponding to the F-value, is then used to decide whether the null hypothesis should be rejected.

Any standard statistical software can be used to perform the calculations. In this laboratory project, we use *MINITAB*. Some basic features of *MINITAB* are outlined in the next section.

SOFTWARE BACKGROUND

When you open *MINITAB*, the window pane is divided into two parts. In the lower part, there is a spreadsheet for entering data, as shown in Figure 7-7. The upper part is where the numerical output from the analyses will appear. All figures that you may request will be displayed in separate windows. The first line of the spreadsheet is reserved for variable names and is not numbered.

After entering the data, one can select among numerous graphing options from the main menu bar under "Graph" and various statistical options from the main menu bar under "Stat." We invite you to enter the data from Table 7-1 and use it to investigate some of the options and to familiarize yourself with what is available. In this project, we shall provide some details relevant to comparing sample means and investigating heritability.

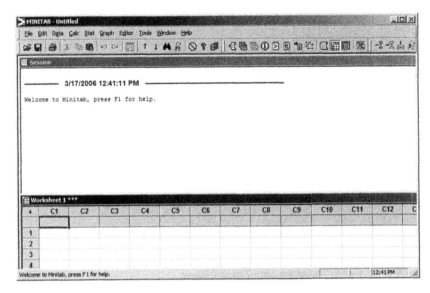

FIGURE 7-7.
Screen shot of *MINITAB*.

Comparing Sample Means. Let's assume that the data from the two samples are entered in separate columns as variables A and B. Assume that we would like to investigate the claim that the mean values between the two groups are different. Then, we would like to see whether we can justify rejecting the claim that the two mean values are the same. To do this, one can choose

"Stat > Basic Statistics > 2-sample t..."

The dialog shown in Figure 7-8 appears: Select "Samples in different columns" and then click in the text field "First." The variables appear in the text area on the left. You can either select your variable by double-clicking it, or you can type the variable name directly. When testing one-sided hypotheses, the order is important, because the null hypothesis will be phrased in terms of the test difference $\mu_{First} - \mu_{Second}$ where μ_{First} and μ_{Second} are the mean values of the random variables in the first and second columns, respectively.

Clicking the "Options..." button brings up the dialog shown in Figure 7-9: The specified "Test difference" of 0.0 corresponds to $\mu_{First} - \mu_{Second} = 0$, which means that your null hypothesis will be H$_0$: $\mu_{First} - \mu_{Second} = 0$. Under "Alternative," there are three choices: Not Equal, Less Than, and Greater Than. If you choose Not Equal, the alternative will be H: $\mu_{First} - \mu_{Second} \neq 0$. Choosing Less Than means that H$_0$: $\mu_{First} - \mu_{Second} < 0$, and so on. After choosing the desired setting, clicking OK closes the dialog. Under "Graphs...," one can select the preferred graphical representation of the data by checking the respective boxes. After all preferences are set, clicking the OK button will perform the specified analyses.

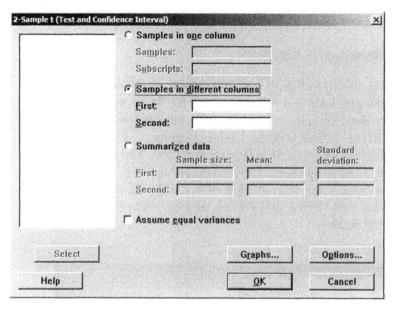

FIGURE 7-8.
MINITAB 2-Sample *t*-test dialog box.

Determining Heritability. In the context above, this resorts to the use of regression analysis. You can obtain a plot of the data with a fitted regression line by either choosing

"Graph > Scatterplot"

and then specifying With Regression, or by selecting

"Stat > Regression > Fitted Line Plot..."

and choosing Linear in the dialog that opens.

We invite you to take some time to investigate these and other graphical options with the data from Table 7-2 (Exercise 7-3) below.

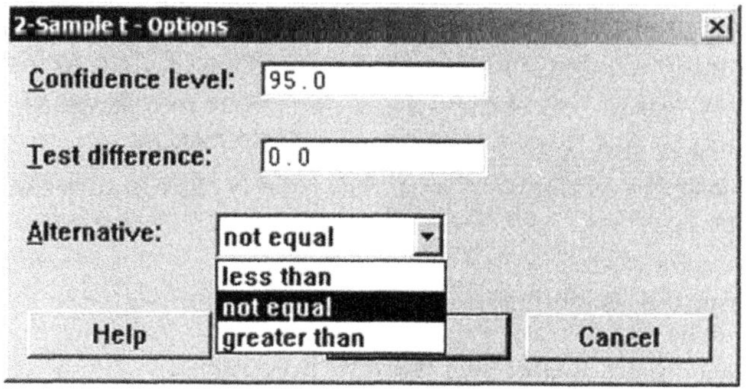

FIGURE 7-9.
MINITAB 2-Sample *t*-test options dialog box.

To perform the analyses, from the main menu, choose

"Stat > Regression > Regression..."

The dialog shown in Figure 7-10 will appear, in which you need to specify the predictor and response variables. There are numerous choices under "Options..." that we will skip for now, and we shall work with the default settings. Under "Graphs..." you are allowed to specify preferences regarding the residual plots. The options under "Results..." and "Storage..." give different options controlling for the display of results and storage of results. Accept the defaults here. When done with setting preferences, clicking OK will display the graphs and results.

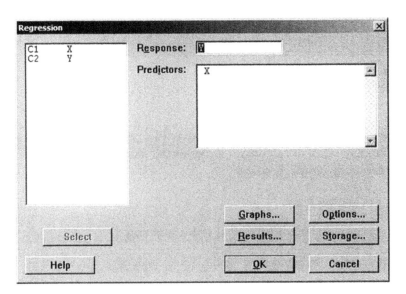

FIGURE 7-10.
MINITAB Regression dialog box.

Plant No.	A (Yield, kg)	Plant No.	B (Yield, kg)
1	2.2	1	2.4
2	2.8	2	2.8
3	1.9	3	3.1
4	3.2	4	2.6
5	2.6	5	2.5
6	2.1	6	2.8
7	2.7	7	3.2
8	2.4	8	3.4
9	2.5	9	2.9
10	2.0	10	2.7

TABLE 7-1.
Data for Exercise 7-1.

Exercise 7-1

An agricultural company is attempting to alter a type of corn to produce a new variety (B) that will be superior in terms of yield to the original variety (A). Table 7-1 contains yield data from two samples drawn from variety A and B, respectively. Use *MINITAB* to quantitatively assess the evidence, if any, that type B corn is superior in terms of yield.

Provide, in writing, the following:

1. The hypothesis being tested and the null hypothesis

2. The statistical test used and justification for using it

3. The *MINITAB* output

4. Interpretation of the output

5. Conclusion

Exercise 7-2

Germination and growth of seeds is affected by many factors: type of seed, type of soil, solar exposure, amount of water, and other environmental factors. Magnetism is a powerful force affecting many physical processes, and it could potentially affect the growth, development, and germination of seeds. Scientist U. J. Pittman concluded, "The roots of some plants [winter and spring wheat, and wild oats] normally align themselves in a North-South plane approximately parallel to the horizontal face of Earth's magnetic field." (Pittman, 1970) If the Earth's own magnetism affects the way plants grow, perhaps added magnetism will affect their rate and quality of growth as well. Experiment 7-1, described below, is designed to explore these questions.

Use the observations/data from Experiment 7-1 described below to formulate one or more hypotheses regarding the effect the presence of a magnetic field may have on the germination of the seeds and the growth of the seedlings. The actual data are recorded in the file *seedlab.xls* which can be downloaded from http://www.biomath.sbc.edu/data.html. Use appropriate statistical techniques to corroborate or reject these hypotheses. Present a clear summary of your findings, including appropriate tables, plots, and charts. Based on your findings, what additional studies would you propose to investigate the stated hypotheses further?

Experiment 7-1

Materials: One hundred lentil seeds; one magnet (from an audio speaker, approximately 7 cm in diameter, 1 cm in height) to provide the magnetic field; two bowls (approximately 19 cm in diameter, 5 cm in height) to serve as flower pots, cotton as a medium for initial germination; water (1 cup every 2 days); and potting soil (2.5 cups per bowl).

Procedure:

1. Lentil seeds were placed on a layer of damp cotton for 24 hours to help speed germination.

2. Fifty seeds were then planted in each bowl, using the same batch and amount of potting soil.

3. One of the bowls was placed on top of the magnet.

4. Bowls were placed 10 feet apart under otherwise similar environmental conditions, including light.

5. The seeds were watered for 2 weeks.

6. The number of seedlings was counted at end of the 2-week experimental period and their heights were recorded.

Observations/Data:

- After 24 hours in the cotton, most of the seeds had opened and had begun to sprout.

- After 2 weeks, 47 of 50 seeds sprouted in the bowl exposed to magnetism.

- After 2 weeks, 43 of 50 seeds sprouted in the bowl not exposed to magnetism.

- The sprouts under the influence of magnetism appeared to be taller than the sprouts not under the influence of magnetism.

Exercise 7-3

Consider Table 7-2, presenting a sample of children's height (stature) and the average parental stature for each child. The question we want to examine is the dependence of child's stature upon the average parental stature. Enter the data in *MINITAB*, and address the questions below.

1. Determine the heritability ratio h^2.

2. Determine whether the result from part (1) is statistically significant.

3. Provide a conclusion based on the data and your answers to parts (1) and (2).

For all of the above, present your findings in writing. Wherever appropriate, specify the hypothesis being tested, the null hypothesis, and the statistical tests being used, together with appropriate justification. Include all relevant *MINITAB* output together with detailed interpretation in the context of this exercise.

Family No.	Average Parental Stature (X, feet)	Child's Stature (Y, feet)
1	5.60	5.70
2	5.90	6.10
3	6.10	6.20
4	5.30	5.60
5	5.70	5.45
6	6.20	5.90
7	6.40	6.10
8	5.50	5.80
9	5.20	5.40
10	6.10	6.20

TABLE 7-2.
Data for Exercise 7-3.

REFERENCES

Bergman, R. N., Zaccaro, D. J., Watanabe, R. M., Haffner, S. M., Saad, M. F., Norris, J. M., Wagenknecht, L. E., Hokanson, J. E., Rotter, J. I., & Rich, S. S. (2003). Minimal model-based insulin sensitivity has greater heritability and a different genetic basis than homeostasis model assessment or fasting insulin. *Diabetes, 52,* 2168–2174.

Devore, J., & Peck, R (2005). *Statistics—The exploration and analysis of data* (5th ed). Belmont, CA: Brooks Cole-Thompson Learning.

Falconer, D. S. (1989). *Introduction to quantitative genetics* (3rd ed). New York: John Wiley & Sons.

Pittman, U. J. (1970). Magnetotropic responses in roots of wild oats. *Canadian Journal of Plant Science, 50,* 350.

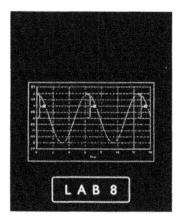

LAB 8

Blood Glucose Fluctuation Characteristics in Type 1 versus Type 2 Diabetes Mellitus

Diabetes mellitus is a complex of disorders, characterized by a common final element of high levels of blood glucose. Diabetes mellitus has two major types: type 1 (T1DM), caused by autoimmune destruction of insulin-producing pancreatic β-cells, and type 2 (T2DM), caused by defective insulin action (insulin resistance) combined with progressive loss of insulin secretion.

Twenty-one million people are currently afflicted by diabetes in the United States, and an ongoing increase in the number of new cases is now assuming epidemic proportions. For example, from 1990 to 1998 the number of adults with diabetes in the United States increased by one third. Because obesity is one of the predisposing factors for diabetes, the high levels of obesity among both children and adults in the United States make it likely that those numbers will increase even more quickly in the near future.

The risks and costs of diabetes (more than $100 billion per year) come from its chronic complications in four major areas: retinal disease, which is the leading cause of adult blindness; renal disease, which represents half of all kidney failures; neuropathy, which predisposes to more than 65,000 amputations each year; and cardiovascular disease, which is two to four times more common in diabetics than in those without diabetes. Cardiovascular disease in diabetes is also more morbid, more lethal, and has benefited less from modern interventions such as bypass surgery.

In a healthy person, the blood glucose (BG) level is internally regulated through insulin release from the pancreas that counterbalances carbohydrate intake. In patients with diabetes mellitus, this internal self-regulation is disrupted. The standard daily control of T1DM involves multiple insulin injections or a continuous insulin infusion (insulin pump) that lowers BG. The control of T2DM may include any combination of diet, exercise, oral medication, or insulin injection.

Large-scale research studies, including the 10-year Diabetes Control and Complications Trial (1993) and a similar European trial (Reichard and Phil [1994]), have proved that intensive treatment with insulin and with oral medication remains the best strategy for optimal glycemic control. Such therapy has been proved effective in bringing blood glucose to nearly normal levels and in markedly reducing the chronic complications of diabetes. However, the same studies have shown that external BG control is still not nearly as good as normal internal self-regulation: too little insulin results in chronic high BG levels (a condition known as *hyperglycemia*), causing complications in multiple body systems over time, whereas too much insulin results in dangerously low BG levels (a condition referred to as *hypoglycemia*). Without corrective action, hypoglycemia can rapidly progress to severe hypoglycemia (SH), a condition identified as low BG resulting in stupor, seizure, or unconsciousness that precludes self-treatment. If the patient does not receive treatment during an SH episode, death can occur. A retrospective population survey from Norway investigating 246 deaths from 1981 to 1990 among diabetic patients younger than 40 years of age attributed 10% of these deaths to hypoglycemia (Bloomgarden [1998]).

Recently, other negative consequences of SH have been documented. Magnetic resonance imaging (MRI) studies revealed cases where occurrences of SH were associated with anatomical changes in the brain, and other studies report permanent cognitive dysfunction. In short, on one side, intensive therapy could lead to improved metabolic control in diabetes; on the other, such therapy was associated with at least a three-fold increase in SH (Diabetes Control and Complications Trial Research Group [1997]). Because SH can result in danger to oneself and others (accidents, coma, and even death), the increased risk for SH discourages patients and health care providers from pursuing intensive therapy. Consequently, hypoglycemia has been identified as the major barrier to improved glycemic control.

Thus, patients with diabetes face a lifelong clinical optimization problem: to maintain strict glycemic control without increasing their risk for hypoglycemia. A mathematical problem associated with such optimization is to create analytical procedures that would continuously assess biologic and behavioral characteristics of hypoglycemia and hyperglycemia. For example, one can envision that over-the-counter blood glucose monitors could serve as a self-monitoring blood glucose (SMBG) data analysis system that is able to identify periods of high risk for hypoglycemia and, during such periods, initiate even higher-level risk analyses, in addition to alerting the patient. The main questions thus become: Is it possible to build such self-monitoring systems? What would be the mathematical tools and models that could solve this problem?

In this laboratory exercise, we shall develop and test a mathematical model designed to assess the risk for SH from blood glucose measurements of people with diabetes collected by home SMBG devices. You will:

- Rescale the data to obtain symmetric samples and thus ensure that certain well-known statistical techniques will be valid.

- Define a risk function that measures the risk of dangerously low and high deviation of the blood glucose from the clinically safe level.

- Test that the new risk function is indeed a superior tool in predicting future episodes of severe hypoglycemia in comparison to traditional measures, such as the glycosylated hemoglobin discribed below.

BIOLOGICAL AND MEDICAL BACKGROUND

It has been well known for more than 20 years that glycosylated hemoglobin (HbA_{1c}) is a marker for the glycemic control of individuals with diabetes. Numerous studies have investigated this relationship and found that glycosylated hemoglobin generally reflects the average BG levels of a patient over the previous 2 months. Because BG levels fluctuate considerably in the majority of patients with diabetes, it has been suggested that the real connection between integrated glucose control and HbA_{1c} can be observed only in patients known to be in stable glucose control over a long period of time. Early studies of such patients have derived an almost deterministic relationship between the average BG level in the preceding 5 weeks and HbA_{1c}—a nearly linear association. In 1993, the Diabetes Control and Complications Trial concluded that HbA_{1c} is "the logical nominee" for a gold-standard glycated hemoglobin assay and established a linear relationship between the preceding mean BG and HbA_{1c}. Guidelines were developed stating that an HbA_{1c} of 7% corresponds to a mean BG of 8.3 mmol/L (150 mg/dl); an HbA_{1c} of 9% corresponds to a mean BG of 11.7 mmol/L (210 mg/dl); and a 1% increase in HbA_{1c} corresponds to an increase in mean BG of 1.7 mmol/L (30 mg/dl).

However, HbA_{1c} has repeatedly proven to be an ineffective assessment of patients' risk for SH. In fact, the Diabetes Control and Complications Trial (1993) concluded that only about 8% of future SH episodes can be predicted from known variables, including HbA_{1c}, and this prediction was only improved to 18% by a recent structural equations model using history of SH, hypoglycemia awareness, and autonomic score. The reason for that poor prediction is quite understandable: HbA_{1c} reflects the average BG level over a few weeks preceding the measurement, but is *not sensitive to the relatively quick and sharp BG transitions in the lower BG range that are responsible for SH*. In other words, HbA_{1c}, although good for long-term assessment of average BG, is too slow a measure to reflect rapid BG changes, even if its continuous monitoring were possible. Given that intensive therapy increases the risk for hypoglycemia, strict control of diabetes implies that BG levels should be closely monitored for large deviations to both the low and the high end of the BG scale. It also follows that the *risk for hypoglycemia needs to be monitored by means other than HbA_{1c}*. The alternative that we consider here uses SMBG data to predict SH.

As we already pointed out, about 10 years ago HbA$_{1c}$ was viewed as the only simple test for patients' glycemic control, and it was believed that measuring mean BG directly was not practical. However, the rapid development of home BG monitoring devices has somewhat changed this conclusion. Contemporary memory meters store up to several hundred SMBG readings and can calculate various statistics, including the mean of these BG readings. Increasing industrial and research efforts are concentrated on the development of devices for continuous, or nearly continuous, more or less invasive monitoring of BG. The meters are usually accompanied by software that has expanded capabilities for data analysis, review, and graphical representation. Two new journals, *Diabetes Technology and Therapeutics* and *Diabetes Science & Technology*, are now devoted to technological advances, including information processing. It is currently believed that diabetes-specific data analysis procedures can substantially improve the forecast of hypoglycemia and the overall quality of the monitoring of diabetes control.

The method for predicting SH described in this laboratory project uses SMBG data combined with recently developed mathematical models and analyses. The original work can be found in Kovatchev et al. (1997).

MATHEMATICAL AND STATISTICAL BACKGROUND

There are no new mathematical concepts necessary for this project. We use standard mathematical techniques for solving algebraic equations and invoke relevant software when necessary to obtain their numerical solutions. As far as statistical techniques, familiarity with applying the *t* test for comparing group averages will be assumed. Details regarding the *t* test and the interpretation of the statistical outcome can found in Chapter 4 of the text and in standard statistical textbooks.

NECESSARY SOFTWARE

Any statistical package or spreadsheet software could be used to perform the data analyses. In addition, a scientific calculator or a computer program capable of determining numerical solutions of algebraic equations will be necessary. In what follows, we use *MS Excel* for both graphical assistance and to facilitate the statistical analyses, but any other comparable software would be appropriate.

PART I: BUILDING A MATHEMATICAL MODEL FOR ASSESSING THE RISK FOR HYPO- AND HYPERGLYCEMIA

The BG Measurement Scale and Its Asymmetry. The level of BG in Système International (SI) units is measured in millimole per liter (mmol/L).

The range of BG in a living human is approximately 1.1 to 33.3 mmol/L, and the safe (target) range is considered to be 3.9 to 10 mmol/L. Values below 3.9 mmol/L form the *hypoglycemic range.* Values above 10 mmol/L form the *hyperglycemic range.* The target (*euglycemic*) range is defined to be from 3.9 mmol/L to 10 mmol/L. The clinically desired BG values of patients with diabetes are those around 6 to 6.5 mmol/L.

When we plot these ranges on a 1- to 33.3-mmol/L scale and mark the hypoglycemic, target, and hyperglycemic ranges (see Figure 8-1), it is apparent that the hyperglycemic range is much wider than the hypoglycemic range and that the target range is not centered within the scale.

As a result, the numerical center of the scale (17 mmol/L) is distant from its "clinical center"—the desired BG values around 6.25 mmol/L. Consequently, when BG readings are analyzed, the assumptions of many parametric statistical methods are not applicable (or are routinely violated). It is thus necessary that any BG data set be *transformed to a symmetric distribution* before using parametric statistics. In addition, for practical and clinical application, it is imperative for the data transformation to be *sample-independent.* This means that transformation should be derived from clinical assumptions, not from a particular BG data set. We next build the desired scale transformation.

Symmetrization of the BG Measurement Scale. Because the BG level is a concentration of sugar in the blood and, as such, it is a continuous function that follows a logarithmic pattern, we seek a scale transformation in the general form

$$f(\text{BG}) = f(BG, \alpha, \beta) = [(\ln(BG))^{\alpha} - \beta], \alpha, \beta > 0,$$

where α and β are positive parameters. Our choice for this form of the transformation was influenced by some classical work (see, for example, Box and Cox [1964]) on correcting skewness in data distributions. The following clinical assumptions should be taken into consideration when determining the values of α and β:

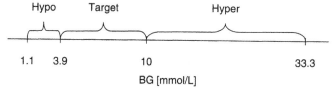

FIGURE 8-1.
BG ranges based on Système International (SI) millimole per liter units.

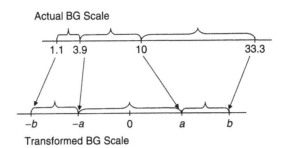

FIGURE 8-2.
Conversion of the BG scale.

1. The directions of the original and transformed scales are the same.

2. The target range is centered at 0.

3. The entire BG range is centered at 0.

In other words, we want to convert the original scale to a scale as shown in Figure 8-2. The idea is to expand the hypoglycemic range, squeeze the hyperglycemic range, and position the target BG range symmetrically about zero.

Finally, by multiplying by a third parameter, γ, we get $f(BG,\alpha,\beta,\gamma) = \gamma [(\ln (BG))^{\alpha} - \beta]$. This third parameter will allow for calibration of the new scale. We require that the following additional condition be satisfied:

4. The transformation should map the minimal and maximal BG values to $-\sqrt{10}$ and $\sqrt{10}$, respectively. This technical condition will be convenient for two reasons. First, a Gaussian random variable with a central normal distribution would have 99.8% of its values between $-\sqrt{10}$ and $\sqrt{10}$, and, second, this will enable us to calibrate the risk function that we will define shortly to be a function with values from 0 to 100.

Exercise 8-1

Express conditions 2 and 3 in terms of the transformation function $f(BG,\alpha,\beta)$.

Exercise 8-2

Express condition 4 in terms of the transformation function $f(BG,\alpha,\beta,\gamma)$.

Exercise 8-3

Solve the equations from Exercises 8-1 and 8-2 to find $\alpha,\beta,\gamma > 0$ for which the transformation $f(BG,\alpha,\beta,\gamma)$ satisfies conditions 2 through 4. Verify that condition 1 is also satisfied.

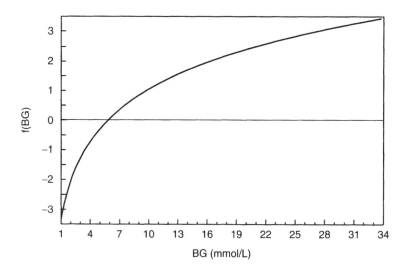

FIGURE 8-3.
The graph of the transformation $f(BG)$ for the values of $\alpha, \beta, \gamma > 0$ determined in Exercise 3.

Hint: Notice that simple algebraic manipulations could easily lead to simplifying the system of equations and give a single equation for α. For this equation, a calculator or a computer program could be used to find a solution for $\alpha > 0$. Back substitutions can then be used to find β and γ.

Alternatively, brute force could be used to find α, β, γ simultaneously by entering the system of equations that corresponds to conditions 2 through 4 into *MATLAB, DERIVE,* or any comparable computer program.

The graph of $f(BG)$ you obtain should look somewhat like the graph in Figure 8-3.

Exercise 8-4

(a) Using the values of α, β, and γ, determine the intervals in the transformed scale over which the transformed hypoglycemic, euglycemic, and hyperglycemic ranges are mapped by the transformation $f(BG)$. **Note:** This is equivalent to determining the unknown values $a, b > 0$ from Figure 8-2.

(b) Determine the BG value for which $f(BG) = 0$.

(c) Fill in the values marked with question marks in Figure 8-4.

Defining a Risk Function and the Low and High BG Indices. We now define a risk function that will assign a risk value to each BG level from 1.1 to 33.3 mmol/L. Figure 8-5 presents a quadratic risk function superimposed over the transformed BG scale.

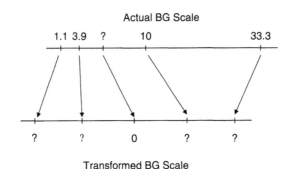

FIGURE 8-4.
Template for answers to Exercise 4(c).

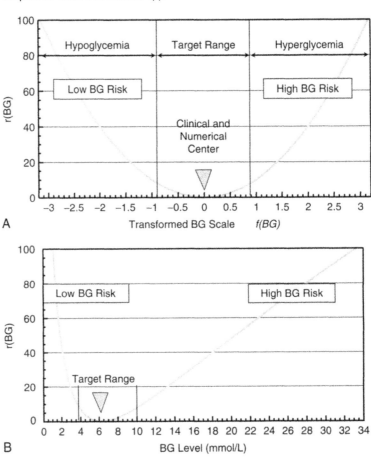

FIGURE 8-5.
The risk function $r(BG)$ on the transformed scale (panel A) and on the original scale (panel B). (From Kovatchev, B.P., Straume, M., Cox, D.J., & Farhy, L.S. [2001]. Risk analysis of blood glucose data: A quantitative approach to optimizing the control of insulin dependent diabetes. *Journal of Theoretical Medicine, 3*, 1-10. Used by permission of Taylor & Francis, Ltd. [http://www.informaworld.com].)

The equation of the BG risk function in Figure 8-5(A) is:

$$r(BG) = 10[f(BG)]^2.$$

We can now better understand the calibration condition imposed on the transformation $f(BG)$ earlier. Because $f(BG)$ ranges from $-\sqrt{10}$ to $\sqrt{10}$, the risk function $r(BG)$ ranges from 0 to 100. Its minimum value of 0 is achieved at $f(BG) = 0$ or, in the original scale, BG = 6.25 mmol/L. The maximum of 100 is reached at the extreme ends of the BG scale, $f(BG) = -\sqrt{10}$ that corresponds to BG = 1.1 mmol/L in the original scale (extreme hypoglycemia) and $f(BG) = \sqrt{10}$ that corresponds to BG = 33.3 mmol/L in the original scale (extreme hyperglycemia). Thus, $r(BG)$ can be interpreted as a measure of the risk associated with a certain BG level. The left branch of this parabola identifies the risk of hypoglycemia, whereas the right branch identifies the risk of hyperglycemia. Notice, again, that because in this scale the hypo- and hyperglycemic ranges of the BG scale are symmetric about 0, the symmetric risk function in Figure 8-5(A) would be equally sensitive to hypoglycemic and to hyperglycemic readings.

For comparison, Figure 8-5(B) presents $r(BG)$ in the original BG scale. As you may have expected, the risk function in this scale increases much more rapidly in the hypoglycemic range and thus is not equally sensitive to hypoglycemic and to hyperglycemic readings.

We next want to use the risk function $r(BG)$ to build a measure (a model) that assesses the risk for hypoglycemia caused by low BG readings. In the same way, we want to have a measure that assesses the risk for hyperglycemia based on the high BG readings. To do this, we separate the low scores (those for which $f(BG) < 0$) from the high scores (those for which $f(BG) > 0$). We will begin with an example and then show the general formula.

EXAMPLE

Suppose the $f(BG)$ readings for a patient are $-1, 2, -0.4, 1, 2.5$. The low readings are thus -1 and -0.4. The risk function values for these are:

$$10[f(BG)]^2 = 10(-1)^2 = 10 \text{ and } 10[f(BG)]^2 = 10(-0.4)^2 = 1.6.$$

We now sum these values and divide by 5 (the total number of readings). Doing this gives

$$\frac{1}{5}(10 + 1.6) = \frac{11.6}{5} = 2.32.$$

This will be the low BG index (LBGI) for these five readings.

Similarly, the high BG index (HBGI) is

$$\frac{1}{5}(10(2)^2 + 10(1)^2 + 10(2.5)^2) = \frac{1}{5}(40 + 10 + 6.25) = \frac{112.5}{5} = 22.5.$$

This example was an illustration for the following general situation. Let $x_1, x_2, \dots x_n$ be n BG readings of a subject, and let

$$rl(BG) = r(BG) \text{ if } f(BG) < 0 \text{ and } 0 \text{ otherwise} \qquad (8\text{-}1)$$

$$rh(BG) = r(BG) \text{ if } f(BG) > 0 \text{ and } 0 \text{ otherwise.} \qquad (8\text{-}2)$$

The LBGI and the HBGI are then defined as:

$$LBGI = \frac{1}{n} \sum_{i=1}^{n} rl(x_i) \qquad (8\text{-}3)$$

$$HBGI = \frac{1}{n} \sum_{i=1}^{n} rh(x_i). \qquad (8\text{-}4)$$

The LBGI is based on the left branch of the BG risk function, whereas the HBGI is based on the right branch of the BG risk function [see Figure 8-5(A)]. The sum of LBGI + HBGI has a theoretical upper limit of 100.

Typically, when the BG readings x_1, x_2, ..., x_n are downloaded from the SMBG device, they will be stored in a data file. The data can then be read into a spreadsheet program or a statistical program for data analysis. In Part II of this laboratory exercise, the data we use for validation of the HBGI and LBGI, defined in Eq. (8-3) and Eq. (8-4), are stored in *MS Excel* spreadsheet format.

For the rest of the section, we review the specific *MS Excel* syntax that will be used. If you are familiar with the IF and AVERAGE functions of *Excel* and know how to use *Excel* to perform a Student t test, you may omit this material and continue with Part II of the laboratory exercise.

If you have never before used these *MS Excel* functions, our brief review may be insufficient. If this is the case, consulting with the *MS Excel* documentation (by performing a search on these functions in the Help menu index) may prove to be helpful. The IF function (necessary for calculating $rl(BG)$ and $rh(BG)$ for every BG reading) in *Excel* has the following format:

IF(logical_test, value_if_true, value_if_false)

It returns one value if the specified logical test is true and another value if it is false. The AVERAGE function (necessary for computing the LBGI and the HBGI) returns the arithmetic mean of its arguments.

To use these functions in *Excel*, when the spreadsheet containing your data is open, use Insert $\rightarrow$ Function from the main menu, and follow the instructions for specifying the functions' arguments.

The TTEST function (necessary for Exercise 7) returns the probability associated with a Student t test (the p-value of the test).

This function has the following syntax:

TTEST(array1,array2,tails,type)

where "array1" is the first data set and "array2" is the second data set. The parameter "tails" specifies the number of distribution tails. If we enter 1 for tails, TTEST uses the one-tailed distribution. If we enter 2 for tails, TTEST uses the two-tailed distribution. Finally, the parameter "type" is the kind of *t* test to perform: 1 stands for a paired *t* test; 2 stands for a two-sample equal variance *t* test; 3 stands for a two-sample different variance *t* test.

If, for example the data for array1 and array 2 are stored on lines 2 through 115 of columns B and C in a spreadsheet,

TTEST (B2:B115, C2:C115, 2,3)

will perform a 2-tailed *t* test under the assumption that the data sets in columns B and C have different variances.

PART II: MODEL VALIDATION

In the next series of exercises, we illustrate the utility of the LBGI and HBGI for evaluation of various aspects of subjects' glycemic control. We begin with a case study comparing data from two subjects with T1DM and T2DM, respectively.

Because of the physiological differences between T1DM and T2DM, patients with T1DM are known to be at much higher risk of experiencing both hyper- and hypoglycemic episodes, and their BG profiles are generally marked with frequent excursions into the hyper- and hypoglycemic BG zones. In contrast, the BG fluctuations of patients with T2DM have smaller amplitudes and span a narrower range about the target zone, because large BG fluctuations are caused by the instability of the insulin–glucose dynamics. A critical factor for such instability is the insulin sensitivity of the body, measured as the amount of glucose that is metabolized per unit of insulin. It is natural to expect that higher insulin sensitivity (or lower insulin resistance) will result in a stronger action of insulin, causing larger BG fluctuations. Because T2DM is a disease of increased insulin resistance, the BG fluctuations in T2DM are less violent.

Our next exercise demonstrates that although both subjects maintained similar average glycemic control as measured by the mean BG value during the period of self-monitoring, the two subjects differ dramatically on other markers of vulnerability to hypoglycemia, as well as in terms of blood glucose irregularity.

For this part, you will need the file *PRES4.xls* that can be downloaded from *http://www.biomath.sbc.edu/data/pres4.xls*.

The file *PRES4.xls* contains two sheets labeled T1DM and T2DM, respectively, containing SMBG data from one T1DM and one T2DM individual, collected over 30 days. The variables included in each data sheet have the following descriptions:

Patient ID	Patient ID Number
Hours	Time of testing since the first test (in hours)
Days	Time of testing since the first test (in days)
BG	BG reading in mmol/L

Exercise 8-5

Use *MS Excel* to create scatter plots of the data with the variable Hours on the horizontal axis and the variable BG on the vertical axis of the BG for each of the data sheets T1DM and T2DM. Comment on the differences you observe.

Exercise 8-6

Use *MS Excel* and the data in *PRES4.xls.* to compute average BG, LBGI and HBGI for the two subjects, T1DM and T2DM, whose data you plotted in Exercise 8-5. Present your results in the table below.

	Average BG	**LBGI**	**HBGI**
T1DM			
T2DM			

Hint: 1. First, create a new column $f(BG)$ that will hold the transformed BG values.

2. Next, create separate columns for $rl(BG)$ and $rh(BG)$, and use *Excel*'s IF function and formulae (8-1) and (8-2) to compute the values for each line in the subjects' files T1DM and T2DM. Your subject files will now contain the following columns.

ID	Hours	Days	BG(mmol/L)	$f(BG)$	$rl(BG)$	$rh(BG)$

3. Finally, use *Excel*'s AVERAGE function to calculate the averages necessary to obtain the values LBGI and HBGI given by Eq. (8-3) and Eq. (8-4).

Exercise 8-7

Use *t* tests to compare these subjects':

(a) Average BGs;

(b) LBGIs; and

(c) HBGIs;

and present the *t* tests' *p*-values in the table below.

	Average BG	**LBGI**	**HBGI**
t test, *p*-value			

Answer the following questions:

1. Did you use one-tail or two-tail *t* tests? Explain your answer.

2. Are the average BG levels for the two subjects significantly different?

3. Are the LBGIs of the two subjects significantly different?

4. Are the HBGIs of the two subjects significantly different?

What conclusions can you make based on these comparisons?

REFERENCES

Bloomgarden, Z. T. (1998). International Diabetes Federation meeting, 1997 and Metropolitan Diabetes Society of New York meeting. *Diabetes Care, 21*, 658–665.

Box, G. E. P., & Cox, D. R. (1964). An analysis of transformations (with discussion). *Journal of the Royal Statistical Society, Series B (Methodological), 26*, 211–252.

Kovatchev, B. P., Cox, D. J., Gonder-Frederick, L. A., & Clarke, W. L. (1997). Symmetrization of the blood glucose measurement scale and its applications. *Diabetes Care, 20*, 1655–1658.

Kovatchev, B. P., Straume, M., Cox, D. J., & Farhy, L. S. (2001). Risk analysis of blood glucose data: A quantitative approach to optimizing the control of insulin dependent diabetes. *Journal of Theoretical Medicine, 3*, 1–10.

The Diabetes Control and Complications Trial Research Group. (1997). Hypoglycemia in the Diabetes Control and Complications Trial. *Diabetes, 46*, 271–286.

The Diabetes Control and Complications Trial Research Group. (1993). The effect of intensive treatment of diabetes on the development and progression of long-term complications of insulin-dependent diabetes mellitus. *New England Journal of Medicine, 329*, 978–986.

FURTHER READING

Reichard, P., & Phil, M. (1994). Mortality and treatment side-effects during long-term intensified conventional insulin treatment in the Stockholm Diabetes Intervention study. *Diabetes, 43*, 313–317.

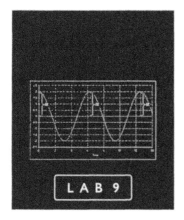

LAB 9

Using Heartbeat Characteristics and Patterns to Predict Sepsis in Neonates

In this laboratory exercise, we consider the question of regularity in data collected over time. We examine this question by considering heart rate data collected from prematurely born infants in an attempt to predict oncoming episodes of a potentially lethal bacterial infection. In our investigation, we examine traditional as well as novel measures for regularity and compare time-independent and time-dependent measures. Statistical techniques are applied to perform validations of the results.

In this lab you will:

- Learn of the health risks associated with premature birth and low birth weight.

- Understand how heart rate is measured and that it varies within certain limits in a healthy organism.

- Examine the link between heart rate patterns and the histogram of the interbeat time intervals.

- Examine how the shape and skewness of this histogram relate to traditional as well as novel statistical characteristics.

- Learn of time-dependent measures for variability, such as sample entropy.

- Test research hypotheses on data collected in the Neonatal Intensive Care Unit of the University of Virginia Hospital.

MEDICAL BACKGROUND

Introduction. Approximately 40,000 very low birth weight (VLBW) infants (i.e., infants with less than 1500 g birth weight) are born in the United States each year (Ventura et al. [1996]). Survival of this group has generally improved with advances in neonatal intensive care in the past decade, but specific diseases, such as late-onset sepsis, continue to be a major cause of morbidity and mortality (Stoll et al. [1996]; Gray et al. [1995]).

The clinical syndrome of sepsis is brought about by the infant's response to insults such as bacterial infection, and it has been named the systemic inflammatory response syndrome (SIRS; Bone et al.

[1997]). According to Stoll et al. (1996), neonatal[1] sepsis occurs in as many as 25% of infants weighing less than 1500 g at birth, and the National Institute of Child Health and Human Development (NICHD) Neonatal Research Network found that neonates who develop late-onset sepsis have a 17% mortality rate (more than twice the 7% mortality rate of noninfected infants of similar weight) and increased morbidity.

Early diagnosis of neonatal sepsis is difficult, because the clinical signs of sepsis are neither uniform nor specific (Escobar [1999]). Although drawing blood and evaluating the culture in a laboratory is considered the gold standard for establishing the diagnosis of sepsis, there are concerns regarding its reliability. This is especially true if single samples of small volume are submitted, as is often the practice in critically ill newborn infants (see Kaftan and Kinney [1998]; Aronson and Bor [1987]; and Kellogg et al. [1997]). The diagnostic challenge is even manifested in the definition given by the Centers for Disease Control and Prevention that allows diagnosis of neonatal "clinical sepsis" with either a negative blood culture or no blood culture at all (see Gladstone et al. [1990]).

These high-stakes uncertainties lead to two shortcomings of current practice in neonatal intensive care units around the country. First, infants with sepsis are often detected only when seriously ill, increasing morbidity and mortality and lowering the chance for prompt, complete recovery with antibiotic therapy. Second, because of the seriousness of the condition, physicians have a very low threshold for suspecting sepsis. Because clinical neonatologists recognize sepsis as a potentially catastrophic illness, they obtain blood cultures and administer antibiotics empirically for subtle symptoms in an attempt to avert disaster. This leads to unnecessary blood cultures and short courses of antibiotics to infants without bacterial infections—in fact, 10 to 20 infants are treated for sepsis for every one infant who has a positive blood culture (Gerdes and Polin [1987]).

Heart Rate Variability. The time elapsed between two sequential heart beats is called the *RR interval*. In health, the time intervals between heartbeats have a certain natural variability, with longer time intervals (decelerations) and shorter time intervals (accelerations) approximately equally present. A useful way to visualize the heart rate variability (HRV) is to plot the length of the RR intervals against the sequential number of every heartbeat, resulting in a plot for the *RR series* observed over a certain period of time.

Figure 9-1(A) represents a typical plot of a healthy RR series. It contains data for a representative 4096-beat series of RR intervals from a single infant, corresponding to approximately 20 to 30 minutes of data. The high points in the plot represent decelerations (i.e., the RR interval ending at this point is longer than the previous one), whereas the low points represent accelerations (i.e., the RR interval ending at this point is

---

1. In medical terminology, infants are referred to as *neonates*.

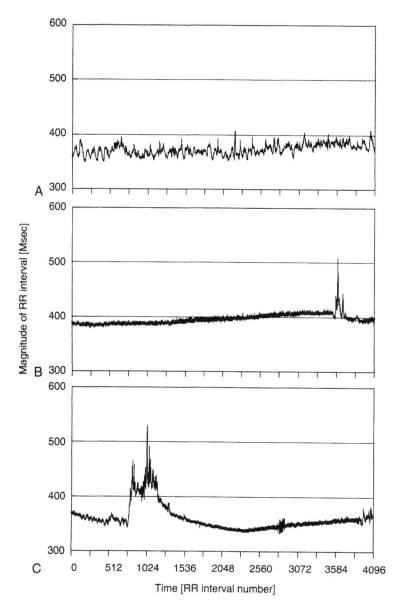

FIGURE 9-1.

Plot of RR interval time series. The three panels are from the same infant while clinically stable (panel A), and from 18 hours (panel B) and 9 hours (panel C) before an acute clinical deterioration leading to death. (From Kovatchev, B. P., Farhy, L.S., Hanquing, C., Griffin, M. P., Lake, D. E., Moorman, J. R. [2003]. *PediatricResearch, 54,* 892–898. Used by permission, © 2003, Pediatric Research, Lippincott, Williams & Wilkins.)

shorter than the RR interval preceding it). The share of accelerations and decelerations is approximately the same. In contrast, in a disease such as sepsis, the plot is visually different when compared with Figure 9-1(A). Figures 9-1(B) and 9-1(C) represent such data. Two questions become important here: (1) What changes in the infants´ health condition are causing the change? and (2) How do we quantify these changes?

A Case Study. The plots in Figure 9-1 are taken from the same infant born prematurely at 24 weeks of gestation with a birth weight of 720 g. The heartbeat data were collected approximately 5 months after birth.

The infant had evidence of chronic lung disease, but was clinically stable around the time the data collection for the data in Figure 9-1 began. However, an acute clinical deterioration in this infant's condition led to the infant's death. The data in panel A were recorded remote from clinical events. The data in panels B and C were recorded 18 and 9 hours before an acute clinical deterioration, and 30 and 21 hours before death, respectively.

The clinical differential diagnosis included septic shock; blood cultures were positive for coagulase-negative *Staphylococcus*, and a urine culture grew *Escheria coli*.

The question remains: Could the heart rate data have been used to save the infant's life, and how could this be done?

In 1995, the NICHD Neonatal Research Network concluded that successful surveillance strategies leading to an earlier diagnosis of sepsis were urgently needed to decrease mortality and morbidity in VLBW infants (see Stoll et al. [1996]). In spite of this long-standing, nationally recognized need, no universally accepted solutions to the problem are yet available. However, recent research at the University of Virginia Medical School and the Wake Forest School of Medicine has established that heart rate characteristics can successfully be used to predict sepsis and SIRS *12 to 24 hours before the clinical diagnosis is made* (see Lake et al. [2002]; Kovatchev et al. [2002]). In this laboratory study, we examine some of the methods used in this research.

MATHEMATICAL AND STATISTICAL BACKGROUND

No new mathematical or statistical concepts are needed for this laboratory project. However, because we explore the relationship between mean, median, and skewness for a body of quantitative data, we shall briefly review the corresponding terminology. If necessary, the reader should review the use of the *t* test for comparing group averages, because this statistical analysis will be utilized repeatedly.

The two most common ways to measure the center of a body of quantitative data are the mean and the median. The mean is what people often refer to as the average and the median is the 50th percentile. In some ways, the median can be more descriptive of the center of the data because it is not affected by extreme values (as the mean can be). A body of data is symmetric if the histogram can be divided into two halves that are mirror images of one another. If the data are symmetric, the mean and the median coincide.[2] In reality, it would be unusual to find a body of data that is symmetric, but many phenomena give data that are approximately symmetric.

---

2. It is possible, however, for the mean and the median to coincide when the data are not symmetric.

Some data that are not symmetric can be described as skewed to the left or right. To get an intuitive idea of skewness, suppose we begin with a symmetric histogram as in Figure 9-2(A). Note that the mean and the median coincide. Now suppose that we take a data point that is to the right of the center and move it farther to the right [Figure 9-2(B)]. The mean moves to the right, but the median says the same. The data in Figure 9-2(B) are skewed to the right.

This example illustrates the following ideas:

1. The mean can be sensitive to extreme values, but the median is less so.

2. If a body of data has a long tail to the right but not to the left, the data is skewed to the right.

3. If the data are skewed to the right, then the mean will lie to the right of the median.

Figure 9-3 presents the general shapes of histograms that depict distribution properties essential for this presentation. In panel A, we have a symmetric distribution. The degree of variability is quantified by the standard deviation (SD)—reduced variability corresponds to a smaller SD. If a sample of RR intervals has predominantly decelerations and few accelerations, its histogram would have a longer right tail and thus its shape will be similar to that in panel B. Conversely, if the sample of RR intervals includes mostly accelerations and few decelerations, the histogram would be similar in shape to that in panel C.

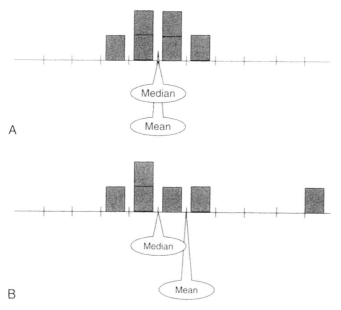

FIGURE 9-2.
A schematic representation of skewness as illustrated by the relative position of the mean versus the median. In panel A, the mean and median coincide, whereas in panel B the mean is to the right of the median.

One way to quantify the balance between the left and right tails of a histogram is to use its *skewness*. The skewness γ of a distribution is defined as the quotient of the third moment μ_3 about the mean $E(X)$ and the third power of the standard deviation σ and given by the expression $\gamma = \dfrac{\mu_3}{\sigma^3} = \dfrac{E\{(X - E(X))^3\}}{[E\{(X - E(X))^2\}]^{3/2}}$. The skewness for symmetric distributions is zero, it is positive if the distribution develops a longer tail to the right of the mean $E(X)$, and it is negative if the distribution develops a longer tail to the left. Consequently, distributions skewed to the right are also called *positively skewed* and distributions skewed to the left are also called *negatively skewed*.

Similar to the standard deviation, however, skewness is based on deviations from the mean value, which is not an accurate measure of the center of an asymmetric distribution. In situations when the interest is mostly in asymmetric distributions, new measures that report on deviations around the median as a more realistic center are needed. The *sample asymmetry* (SA) developed in the following sections is one such measure.

COMPUTER SOFTWARE

Any statistical package or spreadsheet software could be used to perform the data analyses. In what follows, we use *MS Excel* for graphical aid and *MINITAB* to facilitate the statistical analyses, but any other comparable software would be appropriate.

PART I: CONVENTIONAL TIME-INDEPENDENT MEASURES FOR HRV

The data for this project can be downloaded from www.biomath.sbc. edu/data.html. The data set *RR_Data.xls* contains the interbeat intervals data plotted in Figure 9-1. The three data columns in the data file correspond to panels A, B, and C of Figure 9-1, respectively. We shall refer to these columns as data sets A, B, and C.

Exercise 9-1
....

Calculate the mean values for the RR series in data sets A, B, and C. Do you think that the mean value could be used as a good indicator to quantify the abnormalities of the plots B and C? Explain why or why not.

Exercise 9-2
....

Calculate the SDs for the RR data in data sets A, B, and C. Do you think that the SD may be used to identify the abnormal records? Explain why or why not.

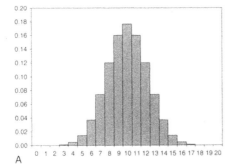

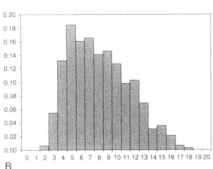

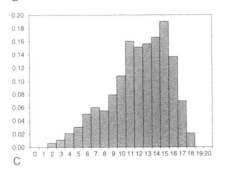

FIGURE 9-3.
Typical histogram shapes of symmetric (panel A), positively skewed (skewed to the right; panel B), and negatively skewed (skewed to the left; panel C) statistical distributions.

Exercise 9-3

(a) Calculate the medians for data sets A, B, and C.

(b) Construct histograms for data sets A, B, and C with the median of each data set positioned at zero. This could be accomplished in the following way: After calculating the median of the RR intervals used in the RR series, the value of the median is subtracted from each data point. The resulting histogram will have exactly the same shape as that of the original data set, but its median will be at zero. The histograms that you will obtain should look similar to those in Figure 9-4.[3]

(c) Do you think that the shape of the histogram could be used as a visual measure for the magnitudes of HRV? Why or why not?

Notice that histograms B and C are positively skewed. In a healthy infant, when approximately equal shares of accelerations and decelerations are present, the histogram will be generally symmetric about the median [Figure 9-4(A)]. The observed abnormality causes marked asymmetry of the histograms in Figures 9-4(B) and 9-4(C), with a reduction of the magnitude of shorter-than-median RR intervals and a clearly right-skewed distribution. This is yet another way to visualize the differences between the normal and abnormal series.

Exercise 9-4

Calculate the skewness for the data sets A, B, and C. Do you think that skewness may be used to identify the abnormal records? Explain your answer.

Note. To obtain the skewness of a data set in MINTAB, from the main menu select
Stats → Basic Statistics → Display Descriptive Statistics...
Click on the "Statistics..." button and select the Skewness checkbox in the dialog that appears. Click OK.

PART II: A NEW TIME-INDEPENDENT MEASURE FOR HRV—SAMPLE ASYMMETRY

We hope that in Part I of this laboratory study you have been able to identify a certain number of limitations to using the SD and skewness of the RR data as measures for dynamic regularity or irregularity. The following observation is a major cause for these limitations. SD and skewness are computed with respect to the mean of the distribution,

---

3. Notice that we have chosen a logarithmic scale for the y-axis.

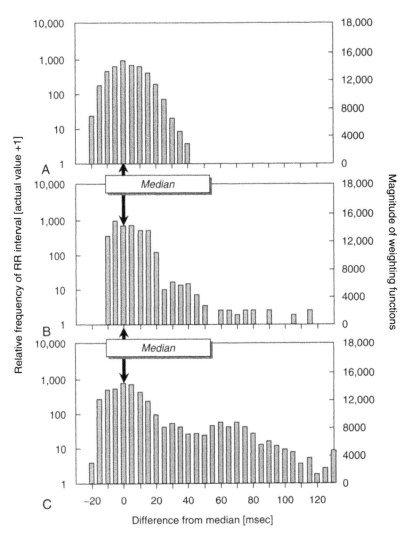

FIGURE 9-4.
Plot of RR interval distribution. The three panels are from the same infant while clinically stable (panel A) and from 18 hours (panel B) and 9 hours (panel C) before an acute clinical deterioration leading to death. (From Kovatchev, B. P., Farhy, L. S., Hanging, C., Griffin, M. P., Lake, D. E., Moorman, J. R. [2003]. *Pediatric Research*, *54*, 892–898. Used by permission, © 2003, Pediatric Research, Lippincott, Williams & Wilkins.)

which is quite vulnerable to large deviations and does not represent accurately the center of a skewed distribution. Figure 9-4, panels B and C, exemplifies this point by presenting samples of RR intervals that include transient decelerations. These decelerations result in an asymmetric distribution with a longer right tail. As a result, the mean and the median of that sample do not coincide—the mean is influenced by a few decelerations and is substantially shifted to the right, describing poorly the center of the RR distribution. In order to overcome this weakness, we need to be able to compute measures that quantify deviations in the length of RR intervals *from their median, or from any other preselected point*. In addition, such measures will need to allow (unlike SD or skewness) for a *separate quantification* of HR accelerations (values less than the median RR interval, forming the left-hand portion of the histogram) from decelerations (values greater than the median RR interval, forming the right). We now introduce the *Sample Asymmetry* (SA), a measure that has these desired properties.[4] That is:

1. SA should grow when there are more decelerations in the RR sequence.

2. SA should decrease as there are more accelerations in the RR sequence.

3. SA should take positive values.

4. SA = 1 for a perfectly symmetric distributions of the RR sequence.

Conditions 3 and 4 are of a technical nature and are, in essence, conditions of calibration. In contrast, conditions 1 and 2 are essential for constructing a measure that overcomes the limitations of the SD and skewness as measures for reduced HRV and transient decelerations.

Conditions 1 and 2 suggest that SA may be defined in the form of a ratio $SA = \dfrac{R_2}{R_1}$, where the numerator R_2 is a measure for the magnitude of RR decelerations and the denominator R_1 measures the magnitude of RR accelerations. Condition 4 will then mean that $R_1 = R_2$; that is, the magnitudes of accelerations and decelerations in the RR sequence are the same—exactly as one would expect from a symmetric distribution. Thus, we can now focus on designing the measures R_1 and R_2.

Let m be the median of the lengths of a sequence of observed RR intervals and x be the length of a single RR interval. For this interval, then, the quantity $(x - m)^2$ could be used as a measure describing the deviation of the RR interval's length from the median length m. In the language we employed in Laboratory 8, we can view this quantity to be the risk assigned to an RR interval of length x because of its deviation from the median value m.

Note also that the problem we faced in Laboratory 8 was quite similar in nature. The question there was to separately assess the risk for hypoglycemia caused by low blood glucose readings and the risk for hyperglycemia caused by high blood glucose readings. Fundamentally, our problem here is identical—we need to assess the magnitude of the risk for RR decelerations, as measured by positive deviations from the median

---

4. This measure was first introduced in Kovatchev et al. (2002).

value, and the risk for RR accelerations, as measured by negative deviations from the median value m. In that sense, R_1 corresponds to the low blood glucose index (LGBI), and R_2 corresponds to the high blood glucose index (HBGI) from Laboratory 8. Thus, following the same approach as in Laboratory 8, for any RR interval of length x, we define the quantities:

$$rd(x) = \begin{cases} (x-m)^2 & \text{for } x > m \\ 0 & \text{for } x \leq m \end{cases}$$

and

$$ra(x) = \begin{cases} (x-m)^2 & \text{for } x < m \\ 0 & \text{for } x \geq m. \end{cases} \tag{9-1}$$

The function $rd(x)$ describes the degree of deviation to the right from the median value (risk for deceleration), whereas the function $ra(x)$ describes the degree of deviation to the left from the median (risk for acceleration).

Consider now a sequence of RR intervals of lengths $x_1, x_2, ..., x_n$, and let m denote the median of the data. Using the weighting functions (9-1), define two quantities representing the sum of the weighting function deviations to the left and to the right from the median μ as follows:

$$R_1 = \frac{1}{n}\sum_n^{i=1} ra(x_i) \quad \text{and} \quad R_2 = \frac{1}{n}\sum_n^{i=1} rd(x_i). \tag{9-2}$$

Definition. The sample asymmetry of the data set $x_1, x_2, ... x_n$ is defined by the ratio

$$SA = \frac{R_2}{R_1}, \tag{9-3}$$

where R_1 and R_2 are the quantities defined by Eqs. (9-2).

In Figure 9-5, we present two examples of data sets—one approximately symmetric (panel A) and one positively skewed (panel B). The graph of the weighing function $(x-m)^2$ is given by a solid black line. As anticipated, the SA for the skewed distribution is higher compared to the SA for the approximately symmetric distribution (3.5 vs 1.1). Notice also that the SA for the data set with an approximately symmetric histogram is close to 1. We leave it to the reader to verify that the SA measure, as defined by Eq. (9-3), satisfies conditions 1 through 4 formulated in the beginning of this section.

To gain a better sense for the capabilities of the newly introduced SA measure, we return to Figure 9-1 and the data set *RR_Data.xls*.

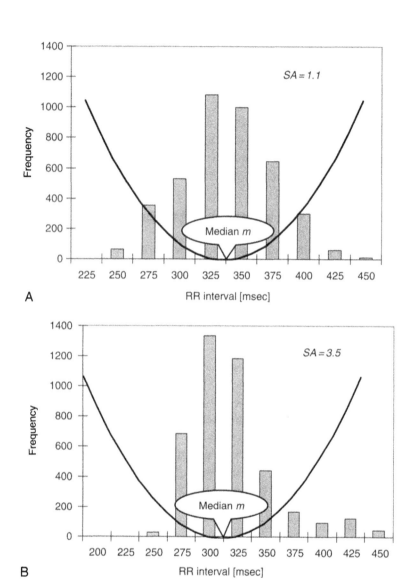

FIGURE 9-5.
SA for an approximately symmetric (panel A) versus nonsymmetric (panel B) distribution of RR intervals.

Exercise 9-5

Calculate R_1, R_2, and SA for the data sets A, B, and C. Compare the results between data sets. What can you say about the change in SA from data set A to data set B and data set C?

We hope that you have observed in the last example that the value of SA increases dramatically for data set C (i.e., before the clinical diagnosis of sepsis and SIRS). You should have also observed that SA was substantially more sensitive to HRV abnormality than the distribution skewness. This is so because R_1 and R_2 can be directly interpreted as representing the frequency and extent of HR accelerations and decelerations, respectively.

The fact that SA gave good results on the specific data sets *RR_Data.xls* we had at hand does not mean that similar encouraging results will be observed on another data set. After all, the change in the SA values we observed with data sets A, B, and C may have been coincidental. To eliminate this possibility, an extensive validation study of the power of SA to predict sepsis and sepsis-like illness in infants needs to be performed before utilizing this measure in clinical settings. We next outline the results of two such clinical validation studies.

SA VALIDATION: TWO RESEARCH STUDIES

The following two studies were designed and undertaken to validate SA as a predictor of upcoming episodes of neonatal sepsis. In both studies, the researchers have demonstrated that abnormal heart rate characteristics (HRC) of reduced variability and transient decelerations occurred early in the course of neonatal sepsis and sepsis-like illness in infants *before the condition had demonstrated itself clinically.*

Study 1: This study introduces a way for quantifying irregularities and changes in the RR series patterns. The measure here, the SA, allows for specific quantification of the contribution of accelerations and decelerations. In the study, 50 infants who experienced a total of 75 episodes of sepsis and SIRS were compared to 50 control infants. The two groups were matched by birth weight and gestational age. RR intervals were recorded for all infants throughout their course in the neonatal intensive care unit at the University of Virginia. It was hypothesized that increases in the SA were indicative of an upcoming episode of sepsis.

The study confirmed that the SA of the RR intervals increased significantly ($p = 0.02$) in the 3 to 4 days preceding sepsis and SIRS, with a steepest increase in the last 24 hours, from a baseline value of 3.3 (SD = 1.6) to 4.2 (SD = 2.3). After treatment and recovery, SA returned to its baseline value of 3.3 (SD = 1.3). Compared to healthy infants, infants who experienced sepsis had similar SA in health and elevated values prior to sepsis and SIRS ($p = 0.002$).

Study 2: This study used 89 infants admitted consecutively to the University of Virginia neonatal intensive care unit over a period of 9 months. A new mathematics index, called *sample entropy* and denoted by SampEn, was developed as a measure for increased or decreased variability in the data RR series. Decrease in the SampEn value is indicative of decreased variability of the RR series (see next section for more details). In contrast, increased SampEn shows that there is more variability in the RR series.

The study confirmed the researchers' hypothesis that SampEn was significantly associated with upcoming sepsis and sepsis-like illness ($P = 0.001$).

Study 1 and Study 2 show that both the SA and the SampEn are good predictors for emergent sepsis episodes and could thus be used as indicators for considering antibiotic treatment before the clinical conditions of sepsis have manifested themselves. The time gained by using these methods (usually between 12 and 24 hours) may be crucial for determining the medical outcomes and saving the infant's life.

PART III: A DYNAMIC MEASURE FOR HRV—SAMPLE ENTROPY

The sample entropy, SampEn, is a way to measure how regular a series of data points is. Heuristically, if the series is perfectly ordered (e.g., the alternating series $-1, 1, -1, 1, -1, 1, \ldots$, and so on), its SampEn is 0. The more deviations from a specific pattern the series exhibits, the higher its SampEn is. Mathematically, however, the situation is more complicated, and the definition depends on several parameters, making use of certain convergence properties. We do not present these details here, but the interested reader will find them in Chapter 6 of the textbook and in Lake et al. (2002). We shall, though, stress once again the empirical use of SampEn in the context of HRV: Reduced variability and transient decelerations in HR data are observed before neonatal sepsis, and SampEn falls before the clinical diagnosis, predicting sepsis by up to 24 hours (see Griffin and Moorman [2001] and Lake et al. [2002]).

A note of caution should be made here. Low SampEn does not distinguish between HR decelerations (which are of interest in predicting sepsis) and HR accelerations (which are not). See Exercise 9-8, where this is discussed in more detail.

Exercise 9-6

Describe why the mean value does not represent a good measure for assessing the variability of the RR series.

Exercise 9-7

Give an example of two data sets that have equal SDs but one of them corresponds to a RR series with transient decelerations while the other one corresponds to one with transient accelerations.

Exercise 9-8

For an RR series with low SampEn, what additional quantitative tests among those discussed in the project and based on the available RR data could be used to determine whether there is an increased share of decelerations (of interest in predicting sepsis) or accelerations? Could you suggest two such tests?

The data sets *GA_BW.xls* and *HRV_DATA.xls* will be used in the Exercises below. The description of the variables recorded in these files is as follows:

ID	Patient ID number
Date	Date of testing
TTE	Time to next sepsis episodes, in days (if TTE = 999.9, there is no next episode)
G. Age	Infant's gestational age
Bwt	Infant's birth weight
Outcome	0 if more than 24 hours before sepsis;
	1 if less than 24 hours before sepsis;
	2 if post-sepsis treatment is under way (these data should be excluded from analysis)
SD	Standard deviation of the HRV distribution
SampEn	Sample entropy computed on 4096 heartbeats
R_1	Left weighting function
R_2	Right weighting function
SampAsm	Sample asymmetry

Exercise 9-9

Use *MS Excel* and the file *HRV_DATA.xls* to create separate time plots of the changes in SD, sample entropy, and SA for Subject 7 from 01/06/00 to 01/14/00. The last four readings of these data are taken within 24 hours before sepsis. Comment on the changes that you detect in the plots.

Exercise 9-10

Use *MINITAB* or another standard statistical software package to:

(a) Import the data spreadsheets.

(b) Use *GA_BW.xls* to compare the gestational age and the birth weight of infants who did versus those who did not experience sepsis during the study (i.e., infants who registered an outcome of 1 versus infants who registered an outcome of 0).

(c) Create box plots for these comparisons to visualize the differences.

(d) Clean *HRV_DATA.xls* by creating a subset containing only outcomes 0 and 1 (2 excluded).

Note. To do this, we need to extract the subset of the data containing only records with value 0 or 1 for the Outcome variable. In *MINITAB*, this could be done as follows:

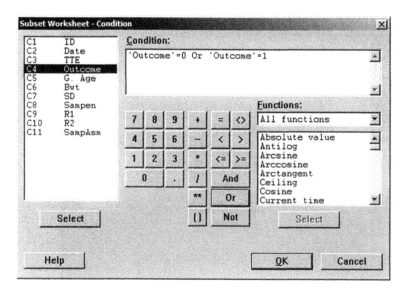

FIGURE 9-6.
MINITAB Subset Worksheet Condition for excluding certain outcomes.

- From the main *MINITAB* menu select "Data → Subset Worksheet."

- Choose the "Specify Which Rows To Include" option under the "Include Or Exclude" selection.

- Click the button "Condition...." A dialog box should appear similar to that shown in Figure 9-6.

- Highlight the Outcome variable in the left panel, and click the Select button.

- Specify 'Outcome' = 0 Or 'Outcome' = 1 as a condition for the selection.

- Click the OK button.

- A new worksheet will now be created that contains only the records with outcomes equal to 0 or 1.

(e) Clean the data by selecting only records with Sample Asymmetry < 100 and only records that precede sepsis (e.g., records for which TTE < 999.9). The latter condition removes all records not followed by an episode of sepsis, and the former excludes records with median near the smallest record values.[5]

---

5. If the median is at or near the smallest record value, R_1 will be close to zero. Thus, records with large values for SA = R_2/R_1 are excluded from the analyses.

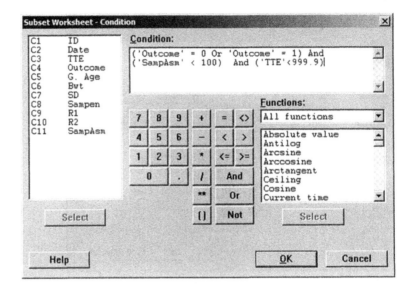

FIGURE 9-7.
MINITAB Subset Worksheet Condition for excluding certain outcomes and SAs.

Note. The entry in the "Condition:" text area should now be as in Figure 9-7.

(f) Using appropriate statistical analyses, compare SD, sample entropy, R_1, R_2, and SA for the times > 24 hours before sepsis versus ≤ 24 hours prior to sepsis. Create box plots for these comparisons to visualize the differences.

Exercise 9-11
....

Using the results from part (f) of Exercise 9-10, list the statistical results supporting the hypotheses that when compared to infants who did not experience sepsis within 24 hours, infants who experienced sepsis had:

(a) Significantly lower birth weight and gestational age;

(b) Significantly lower sample entropy before sepsis;

(c) Significantly higher SA; and

(d) Approximately the same SD, because SD is a marker for sepsis only in a few infants.

Provide the software outputs from the statistical analyses and justify your conclusions.

REFERENCES

Aronson, M. D., & Bor, D. H. (1987). Blood cultures. *Annals of Internal Medicine, 106,* 246–253.

Bone, R. C., Grodzin, C. J., & Balk, R. A. (1997). Sepsis: A new hypothesis for pathogenesis of the disease process. *Chest, 112,* 235–243.

Escobar, G. J. (1999). The neonatal "sepsis work-up": Personal reflections on the development of an evidence-based approach toward newborn infections in a managed care organization. *Pediatrics, 103,* 360–373.

Gerdes, J. S., & Polin, R. A. (1987). Sepsis screen in neonates with evaluation of plasma fibronectin. *Pediatric Infectious Diseases Journal, 6,* 443–446.

Kaftan, H., & Kinney, J. S. (1998). Early onset neonatal bacterial infections. *Seminars in Perinatology, 22,* 15–24.

Gladstone, I. M., Ehrenkrantz, R. A., Edberg, S. C., & Baltimore, R. S. (1990). A ten-year review of neonatal sepsis and comparison with the previous fifty year experience. *Pediatric Infectious Diseases Journal, 9,* 819–825.

Gray, J. E., Richardson, D. K., McCormick, M. C., & Goldmann, D. A. (1995). Coagulase-negative Staphylococcal bacteremia among very low birth weight infants: Relation to admission illness severity, resource use, and outcome. *Pediatrics, 95,* 225–230.

Kellogg, J. A., Ferrentino, F. L., Goodstein, M. H., Liss, J., Shapiro, S. L., & Bankert, D. A. (1997). Frequency of low level bacteremia in infants from birth to two months of age. *Pediatric Infectious Diseases Journal, 16,* 381–385.

Kovatchev, B. P., Farhy, L. S., Hanqing, C., Griffin, M. P., Lake, D. E., & Moorman, J. R. (2003). Sample asymmetry analysis of heart rate characteristics with appilication to neonatal sepsis and systemic inflammatory response syndrome. *Pediatric Research, 54,* 892–898.

Lake, D. E., Richman, J. S., Griffin, M. P., & Moorman, J. R. (2002). Sample entropy analysis of neonatal heart rate variability. *American Journal of Physiology, 283,* R789–R797.

Stoll, B. J., Gordon, T., Korones, S. B., Shankaran, S., Tyson, J. E., & Bauer, C. R. (1996). Late-onset sepsis in very low birth weight neonates: A report from the National Institute of Child Health and Human Development Neonatal Research Network. *Journal of Pediatrics, 129,* 63–71.

Ventura, S. J., Martin, J. A., Mathews, T. J., & Clarke, S. C. (1996). Advance report of final natality statistics. *Monthly Vital Statistics Report, 44,* 1–88.

FURTHER-READING

Griffin, M. P., & Moorman, J. R. (2001). Toward the early diagnosis of neonatal sepsis and sepsis-like illness using novel heart rate analysis. *Pediatrics, 107,* 97–104.

Moro, M. L., DeToni, A., Stolfi, I., Carrieri, M. P., Braga, M., & Zunin, C. (1996). Risk factors for nosocomial sepsis in newborn infants and intermediate care units. *European Journal of Pediatrics, 155,* 315–322.

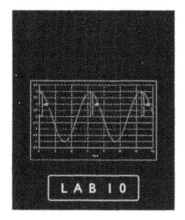

LAB 10

Hormone Pulsatility in Reproductive Endocrinology

In this laboratory project, we examine hormone secretion profiles and apply quantitative methods for determining important features of these profiles such as location, frequency, and magnitude of secretion bursts. These factors are critical with regard to the proper functioning of the endocrine system. Here, we examine the secretion profiles of luteinizing hormone (LH) and the effect that deviations outside the range of normal variability may have on female fertility.

In this laboratory exercise, you will:

- Evaluate the available analysis programs using a simulated data set.

- Examine normal LH data sets to determine:

  (a) The characteristics of normal LH pulsatility patterns (pulse strength and frequency);

  (b) Variation in normal patterns (by comparing data sets from several normal women); and

  (c) How the pattern of LH release varies throughout the menstrual cycle.

- Examine data from infertile patients and infer hypotheses regarding the factors that may be causing infertility.

BIOLOGICAL BACKGROUND

Introduction. Open almost any first-year biology text, and you are likely to find a figure similar to Figure 10-1. It shows the blood serum levels of the pituitary hormones follicle-stimulating hormone (FSH) and LH, of the ovarian hormones estrogen and progesterone, and of the events in the ovaries and uterine lining over a 28-day normal menstrual cycle. Unfortunately, it gives the impression that hormone levels rise and fall smoothly on a time scale of days, which is not the case. In fact, if a woman's LH and FSH levels did behave as shown in the figure, she would be infertile.

Instead, pituitary hormones including growth hormone, prolactin, thyrotropin, adrenocorticotropic hormone, FSH, and LH have been shown to be produced in a pulsatile manner. These hormone levels rise and fall multiple times per day, because of bursts of secretion by

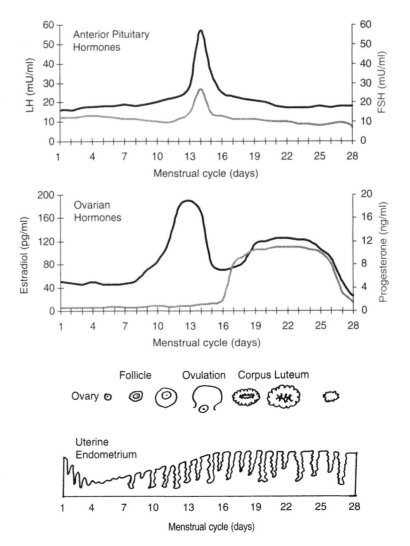

FIGURE 10-1.
Human menstrual cycle, showing hormone levels, response of ovarian follicle, and uterine lining.

the pituitary followed by periods of secretory inactivity (Veldhuis et al. [1987]). The number of secretory bursts per day varies with the hormone in question. It may vary with the age, gender, and health of the individual, and does fluctuate with the phase of the menstrual cycle in women of reproductive age.

Figure 10-2 shows an example of the variation of serum LH levels over 24 hours in a healthy woman of reproductive age. The levels of LH rise and fall multiple times over the course of a single day.

Why does the pituitary produce these hormones in a pulsatile manner? It does so in response to releasing hormones produced by the hypothalamus, which are themselves produced in a pulsatile manner. The hypothalamus is a part of the brain which is located on the lower surface of the brain adjacent to the pituitary. It forms a

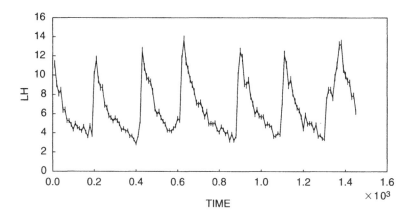

FIGURE 10-2.
Pulsatility of serum LH levels over 24 hours. The data were obtained from a 33-year-old woman during the mid-luteal phase. Time is in minutes; LH is in milliIU (International Units) per milliliter; and the sampling interval was 10 minutes.

coordinating link between the central nervous system and the endocrine system. As a neuroendocrine tissue, the hypothalamus contains neurosecretory cells which produce hormones and releasing factors.

The posterior part of the pituitary gland is an extension of the hypothalamus—the hormones of the posterior pituitary are produced in the hypothalamus and delivered via neurosecretory cells to the posterior pituitary. The anterior part of the pituitary is an endocrine gland whose secretion patterns are controlled by releasing factors or inhibiting hormones secreted into the bloodstream by the hypothalamus. These releasing or inhibiting hormones travel the short distance from the hypothalamus to the anterior pituitary, where they are taken up by surface receptors on the anterior pituitary cells and cause the production of specific hormones. In this case, gonadotropin releasing hormone (GnRH) causes the production of LH and FSH by the anterior pituitary. These hormones stimulate the production of estrogen and progesterone by the ovaries, which in turn control the production of GnRH by the hypothalamus. Thus, the hypothalamus, the pituitary, and the ovaries form a neuroendocrine axis in which production of hormones is controlled by feedback loops.

LH Pulsatility. The gonadotropins LH and FSH are vital to proper menstrual function. LH, together with FSH, stimulates the growth of the ovarian follicle. A mid-cycle peak of LH triggers ovulation—the release of the ovum from the ovary. This peak is clear in Figure 10-1. What is not seen in Figure 10-1, though, is the pulsatile nature of LH secretion, reflecting multiple GnRH pulses per day produced by the hypothalamus. In their work on LH pulsatility, Sollenberger et al. (1990a) went further to show that when pulses of GnRH were administered, LH pulses were generated in response. This demonstrated that the pituitary could be induced to produce LH in an appropriately pulsatile fashion under the influence of exogenously administered GnRH.

Phase	Reported Number of Pulses per 24 Hours
Early follicular	14–24
Late follicular	17–29
Mid-luteal	4–16

TABLE 10-1.
Range of LH Pulses for Three Different Phases of the Menstrual Cycle, as Reported by Sollenberger et al. (1990b).

The number of pulses per day varies over the course of the menstrual cycle. (Figure 10-2 shows one example from the mid-luteal phase of the cycle.) Table 10-1 shows the range of LH pulses reported in the literature for three different phases of the menstrual cycle. The early follicular phase spans the first few days of a new cycle; the late follicular phase corresponds to the mid-cycle (around days 11, 12, and 13); and the mid-luteal phase corresponds to the middle of the 14 days after ovulation (around day 21). Individual variation aside, two possible factors contribute to the huge variation in results: the sampling paradigm (specifically, how often the samples are collected) and the pulse detection algorithm (how the data are analyzed).

Sampling Paradigm. A proper sampling protocol is required because the sampling paradigm determines how many pulses are detected. Simply put, the more often one samples, the more peaks may be detected. For example, examine the following two graphs (Figures 10-3 and 10-4).

The shorter sampling interval of Figure 10-4 allows one to detect more pulses. If the sampling interval were even shorter, it is likely that additional pulses would be detected. A long sampling interval probably explains the smooth curves in Figure 10-1. For example, Veldhuis and Johnson (1986) measured LH levels in 5 men over the course of 8 hours. Sampling at 4-minute intervals gave the greatest number of signals (10.2 ± 2.0), whereas sampling every 32 minutes detected only 2.0 ± 0.0 signals. However, sampling at shorter intervals introduces practical problems in obtaining the data points with a corresponding new set of uncertainties. A compromise seems to lie in a 10-minute sampling interval, which has come to be the standard protocol used

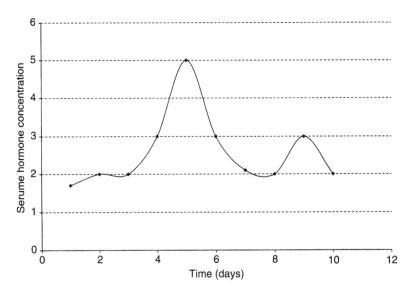

FIGURE 10-3.
Simulated variation of hormone levels over time, with a data collection interval of 24 hours.

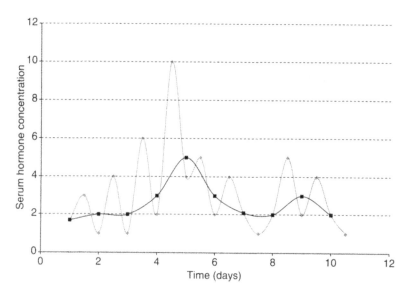

FIGURE 10-4.
Simulated variation of hormone levels over time, with a data collection interval of 12 hours (gray line) and 24 hours (black line).

for LH measurements in the Division of Endocrinology at the University of Virginia School of Medicine (Sollenberger et al. [1990a]; South et al. [1993]).

MATHEMATICAL BACKGROUND

The concentration of hormone in the bloodstream is a consequence of two competing processes—the secretion of the hormone into the bloodstream and its physiological removal from the bloodstream (see Figure 10-5). The structure of a peak in the plasma concentration $C(t)$ derives from both of these processes—the secretory event described by the rate of secretion $S(t)$ into the blood and the physiological elimination (decay) of the hormone $E(t)$ from the bloodstream. Any basal concentration or constitutive expression would also have to be considered.

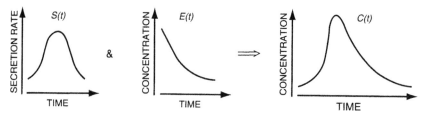

FIGURE 10-5.
A graphical depiction of the convolution process. It shows how the concentration of a hormone in the bloodstream would change over time as a consequence of a single secretory event followed by the removal of the hormone from the bloodstream. (From Veldhuis, J. D., Carlson, M. L., & Johnson, M. L. [1987]. *Proceedings of the National Academy of Sciences of the United States of America* 84, 7686–7690.)

The frequency and amplitude of the hormonal pulses control the reproductive physiology, so an understanding of hormone pulsatility should lead to an understanding of the processes of reproduction. In order to understand the processes, it is important to detect the number of LH secretory episodes, their location, frequency, and duration, as well as their corresponding amplitudes, while also determining the half-life of the LH. How might we best determine the quality and quantity of the hormone pulses?

According to Johnson and Veldhuis (1995), the concentration of a hormone in the serum may be expressed as the convolution integral

$$C(t) = C(0)E(t) + \int_0^t S(z)E(t-z)dz = \int_{-\infty}^t S(z)E(t-z)dz, \ t > 0 \quad (10\text{-}1)$$

where $C_0 = C(0)$ is the concentration of hormone at $t = 0$. Because the concentration in the bloodstream can be measured, and because, in most cases, the elimination function is known, the question is how to obtain the secretion pattern represented by the function $S(z)$. Thus, in order to extract information about the secretory events from the serum hormone concentration data, one must separate the elimination function from the serum concentration profile—in other words, one must perform a *deconvolution analysis*. In the exercises that follow, we shall use software to perform this analysis.

Deconvolution analysis allows for the determination of the frequency and amplitude of the pulses, allowing us to understand the processes controlling normal menstrual function. Using deconvolution analysis, Sollenberger et al. (1990b) determined that the mean number of LH peaks was 17.5 ± 1.4 during the early follicular phase, 26.9 ± 1.6 during the late follicular phase, and 10.1 ± 1.0 during the mid-luteal phase. For each of the phases, this study obtained data from eight women with normal menstrual function and used a sampling interval of 10 minutes for 24 hours for each investigation. Twenty women participated overall, four of whom were studied twice (in two different phases of the cycle).

Once normal function of the hypothalamus–pituitary interaction was understood, it became possible to use deconvolution analysis to diagnose defects of LH production. The result of this research is tangible: Women who are experiencing infertility because of hypothalamic amenorrhea or insulin-dependent diabetes mellitus (South et al. [1993]) may be treated with GnRH to induce appropriate LH production, and women with polycystic ovary syndrome may also be treated to regularize their inappropriate LH production. Successful pregnancies have resulted following this treatment (William S. Evans, MD, personal communication).

COMPUTER SOFTWARE BACKGROUND

In the exercises that follow, you will be using programs in the *PULSE_XP* hormone pulsatility analysis software developed by Dr. Michael Johnson of the University of Virginia. Free download is available from http://mljohnson.pharm.virginia.edu/home.html. To run the programs, once the installation is complete, click on the "pulse_xp" icon or look for "pulse_xp" on the Programs menu. Select the program you wish to run. Follow the instructions for loading data: click on "File," and then "Read a Data File." All of the data files you will use are in ".fix" format, so choose "FIX Mode." As a general rule, begin by accepting the defaults in each of the programs. For each trial that you run, record the initial conditions and any salient features of the results. For a quick tutorial to the package, go to "Help → Help for Hormone" and select the file *Hormone Overview_0601.pdf*.

PART I: USING SOFTWARE TO ANALYZE HORMONE SECRETION PATTERNS

In this section, you will explore the capabilities of the pulse detection software and learn how to interpret the output. In some cases, you may want to print both the graphical and text outputs. Those differ depending on the specific algorithm, but you can expect to see some or all of the following:

1. A graph of hormone concentration with time;

2. The residuals [between the calculated curve (red) and the data points];

3. A representation of secretion events;

4. A representation of autocorrelation;

5. Number of peaks found in the data; and

6. The estimated half-life (HL) of the hormone.

Exercise 10-1

Using each of the indicated pulse analysis algorithms in turn, evaluate the data file *figure1.fix*. For each program, report the number of pulses and half-life determined (if applicable).

1. Produce and submit a table of results similar in format to Table 10-2.

2. What can you conclude from these results? Be sure to consider all of the information the program gives you. Look at the residuals. For example, are they randomly distributed? Does the plot fit the data?

Program Used	Number of Peaks Found	Comments
CLUSTER8		
PULSE		
PULSE2		
PULSE4		

TABLE 10–2.
Sample Table for Presenting the Results from Exercise 10-1.

3. Why do the different algorithms give different results?

4. Which program would you use for analysis of real data, and why?

Exercise 10-2

What does a normal LH secretion pattern look like? Here we examine three data files from a woman with normal menstrual cycles, each file representing serum collection over the course of 12 hours. Three data files, representing the early follicular (*CLHEF1.fix*), late follicular (*CLHLF1.fix*), and mid-luteal (*CLHML1.fix*) phases of the menstrual cycle are found in the data folder. Examine each of these data files using the program you selected after performing Exercise 10-1 above.

How do we interpret the results of the analysis? First, examine the number of peaks in each phase. Then look at the secretion graph and examine the secretion levels. These two values—the number of peaks and the level of secretion—will allow you to make some conclusions about the nature of LH pulsatility in normal women of reproductive age.

1. Report the results of your analyses of *CLHEF1.fix*, *CLHLF1.fix*, and *CLHML1.fix*.

2. What seems to happen to LH secretion over the course of the menstrual cycle?

Exercise 10-3

The files examined in Exercise 10-2 all came from the same woman. Now we shall examine files from additional women with normal menstrual cycles. The files *CLHEF2* and *CLHEF3* are early follicular files; *CLHLF4* and *CLHLF5* are late follicular files; and *CLHML6* and *CLHML7* are mid-luteal files. Examine each of these files and compare to the appropriate results from Exercise 10-2.

1. Comparing the new files to those examined in Exercise 10-2, do you observe any variation in peak number or peak height? (Compare the results for each phase—the new early follicular files to the early follicular file from Exercise 10-2, and so on.)

2. Now that you have a little more data, do your conclusions from Exercise 10-2 still hold? Explain.

CASE STUDIES

Exercise 10-4

Mary Smith, age 34, was referred to the Reproductive Endocrinology Clinic by her gynecologist. Mary and her husband have been married for 10 years and have not been using any contraceptives for the last 5 years. Though initially she was not worried by their failure to conceive, she is beginning to be alarmed. In addition, for the last several years she has experienced fewer and fewer periods. She has taken six over-the-counter pregnancy tests over the last 18 months, all of which have been negative. She finally paid a long-overdue visit to her gynecologist, who promptly made the referral to the clinic.

Mary reported her medical history as follows: Her periods began when she was 12, and she experienced normal menstrual cycles for the next 12 years. At age 24, she began to take oral contraceptives and used them for 5 years. Her cycles resumed normally, though shortly thereafter they seemed to lengthen to 35 days, and over the last few years she thinks they have been down to about four per year. She has had no other unusual symptoms, so there is no evidence of tumors of the brain, thyroid, adrenal glands, or other organs. She follows a regular schedule of moderate exercise and has a normal diet, such that her weight has remained constant over the last 6 years.

The clinician's examination revealed that Mary had normal thyroid function and normal androgen levels. Her pituitary, uterus, and ovaries responded normally when challenged with the appropriate signals. This process of elimination led the clinician to suspect the hypothalamus, and serum LH data were collected.

Examine the data file *PLH11.fix* containing the LH data for this patient. Analyze the data as above. How many peaks are present? What is the height of the peaks? How does this data set differ from the normal cases?

Exercise 10-5

Files *PLH12* and *PLH13* come from patients with medical histories similar to that of Mary Smith. Analyze the data sets as above and compare to the normal files and the *PLH11.fix* file from patient Smith. If these two new patients do indeed have the same condition, how would you describe the characteristics of LH behavior that would be consistent with this condition?

Exercise 10-6

Jane Doe is 23 years old. She has been married for 2 years, and she and her husband have never used contraception, because they are really looking forward to starting a family. Unfortunately, Jane has been unable to conceive. Her husband has been to see a urologist, and his sperm count is normal, so Jane has finally made an appointment with a gynecologist. The gynecologist takes a complete medical history and learns the following facts.

Jane has always had a problem maintaining her weight. She reports that puberty occurred "a few years later" than her peers, and that she had her first period at age 14. She has never had normal cycles, though, and she averages six periods per year. Upon physical examination, the gynecologist notes that Jane has some coarse facial hair, oily skin, and acne. When questioned, Jane admits to plucking the coarsest hairs from her chin and upper lip and to bleaching the areas to make the remaining hair less noticeable. The gynecologist notes some hair on Jane's upper back, and also notes that Jane has a male pattern of pubic hair. She suspects that Jane may be producing an abnormal amount of androgens, and refers her to the Reproductive Endocrinology Clinic.

Evaluation shows that Jane's thyroid hormone levels are normal and her pituitary responds normally. However, her androgen levels are indeed high, and her uterine and ovarian functions are abnormal. The clinician decided to examine hypothalamic function to determine the cause of the symptoms, and LH data were collected.

Examine the data file *PLH21.fix* containing the LH data for this patient. Analyze the data as above. How many peaks are present? What is the height of the peaks? How does this data set differ from the normal cases?

Exercise 10-7

Files *PLH22* and *PLH23* come from patients with medical histories similar to Jane Doe. Analyze the data sets as above and compare to the normal patient files and the *PLH21.fix* file from patient Doe. If these two new patients do indeed have the same condition, how would you describe the characteristics of LH behavior which would be consistent with this condition?

PART II: REVISITING THE EQUATION

$$C(t) = \int_{-\infty}^{t} S(z)E(t-z)dz^1$$

We now examine more closely the equation

$C(t) = \int_{-\infty}^{t} S(z)E(t-z)dz$ and justify this representation of the

concentration function $C(t)$. We begin by recalling that the rate of change in the hormone concentration can be expressed as

$$\frac{dC(t)}{dt} = -kC(t) + S(t), \qquad (10\text{-}2)$$

where k is the elimination rate constant of the hormone (*half-life* $= \ln(2)/k$).

---

1. This part could be included or omitted, depending on the student's proficiency in calculus.

The term $-kC(t)$ is the rate of change of the concentration $C(t)$ due to removal of the hormone from the bloodstream, and the term $S(t)$ represents the rate of change in $C(t)$ because of secretion.

As the next exercise shows, under the assumption that $E(t) = e^{-kt}$ (i.e., assuming exponential removal of the hormone from the bloodstream) the convolution integral

$$C(t) = \int_{-\infty}^{t} S(z)E(t-z)dz$$

is a solution of the differential Eq. (10-2). That is to say, Eqs.(10-1) and (10-2) mean the same thing, written in two different ways.

Exercise 10-8

Prove that if $C(t) = \int_{-\infty}^{t} S(z)e^{-k(t-z)}dz$, then $\dfrac{dC(t)}{dt} = -kC(t) + S(t)$.

Hint: Follow the outline below:

1. Calculate that $C(t+h) = e^{-kh}\int_{-\infty}^{t+h}S(z)e^{-k(t-z)}dz$.

2. Show that
$$C(t+h) - C(t) = (e^{-kh} - 1)\int_{-\infty}^{t}S(z)e^{-k(t-z)}dz + e^{-kh}\int_{t}^{t+h}S(z)e^{-k(t-z)}dz.$$

3. Show that $\dfrac{C(t+h)-C(t)}{h} = \dfrac{e^{-kh}-1}{h}C(t) + \dfrac{e^{-kh}}{h}\int_{t}^{t+h}S(z)e^{-k(t-z)}dz$.

4. Finally, to prove that $\dfrac{dC(t)}{dt} = -kC(t) + S(t)$, use that
$$\lim_{h\to 0}\frac{1}{h}\int_{t}^{t+h}S(z)e^{-k(t-z)}dz = S(t), \text{ and } \lim_{h\to 0}\frac{e^{-kh}-1}{h} = -k.$$

REFERENCES

Johnson, M. L., & Veldhuis, J. D. (1995). Evolution of deconvolution analysis as a hormone pulse detection method. *Methods in Neuroscience, 28,* 1–24.

South, S. A., Asplin, C. A., Carlsen, E. C., Booth, R. A., Weltman, J. Y., Johnson, M. J., Veldhuis, J. D., & Evans, W. S. (1993). Alterations in luteinizing hormone secretory activity in women with insulin-dependent diabetes mellitus and secondary amenorrhea. *Journal of Clinical Endocrinology and Metabolism, 76,* 1048–1053.

Veldhuis, J. D., Carlson, M. L., & Johnson, M. L. (1987). The pituitary gland secretes in bursts: Appraising the nature of glandular secretory impulses by simultaneous multiple-parameter deconvolution of plasma hormone concentrations. *Proceedings of the National Academy of Sciences of the United States of America, 84,* 7686–7690.

Veldhuis, J. D., & Johnson, M. L. (1986). Cluster analysis: A simple, versatile, and robust algorithm for endocrine pulse detection. *American Journal of Physiology, 250,* E486–E493.

FURTHER READING

Sollenberger, M. J., Carlsen, E. C., Booth, R. A., Johnson, M. J., Veldhuis, J. D., & Evans, W. S. (1990a). Nature of gonadotropin-releasing hormone self-priming of luteinizing hormone secretion during the normal menstrual cycle. *American Journal of Obstetrics and Gynecology, 16,* 1529–1534.

Sollenberger, M. J., Carlsen, E. C., Johnson, M. J., Veldhuis, J. D., & Evans, W. S. (1990b). Specific physiological regulation of luteinizing hormone secretory events throughout the human menstrual cycle: New insights into the pulsatile mode of gonadotropin release. *Journal of Neuroendocrinology, 2,* 845–852.

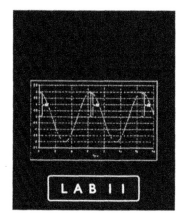

Endocrine Oscillators: Modeling and Analysis of the Growth Hormone Network

In this laboratory exercise, we explore ways of creating dynamical models that represent hormone interactions described schematically by formal endocrine networks. Because one of the most biologically important and intriguing properties of various hormones is the pulsatile pattern of their release, we examine the specific features of the models that generate oscillations in the hormone concentration profiles. We create a mathematical model describing the secretion dynamics of growth hormone (GH) and examine the capabilities of the model to produce hormone concentration profiles similar to those observed experimentally. We also use the model to determine the effect of adding a putative link to the consensus GH network. This demonstrates the value of a mathematical model as an experimental tool for initial examination of hypotheses suggested by experimental observations. More specifically, in this laboratory exercise we examine:

- Symbolic-scheme representations of endocrine models and corresponding mathematical descriptions;

- Delay and feedback factors;

- Modeling hormone (feedback) interactions by means of control functions;

- Approximation of up- and down-regulation by means of Hill functions; and

- Model capabilities of generating and sustaining oscillations.

BIOLOGY AND PHYSIOLOGICAL BACKGROUND

A. Oscillating Systems

Introduction. In *What is a Biological Oscillator?*, Friesen and Block wrote:

"There can be little doubt that oscillations are an essential property of living systems. From primitive bacteria to the most sophisticated life forms, rhythmicity plays a vital role in providing for intercellular communication, locomotion, and behavioral regulation. Although the presence of biological rhythms has been recognized since antiquity, only recently has the origin of these rhythms been systematically addressed. At present, there are numerous descriptions of biochemical, biophysical, and physiological oscillations in the

scientific literature. . . . Most recently, mathematical analysis has been applied to biological oscillators as well". . . .

In view of the extensive mathematical treatment of biological oscillations, we find it curious that much of the experimental literature on biological rhythms contains little or no substantive information on the mechanisms actually responsible for the biological oscillations. . . . This problem exists, we believe, because many physiologists. . . lack a useful formalism with which to describe oscillators. Mathematicians, for their part, have the quantitative procedures to analyze feedback systems, but physiologists often lack the training to utilize effectively these procedures." (Friesen and Block [1984])[1]

Following the formalism introduced by Friesen and Block, in this laboratory exercise we shall focus on three fundamental questions:

1. What are the biological variables essential to the oscillator?

2. How do these essential variables interact?

3. Can these interactions lead to oscillations?

To be able to answer these questions, we shall examine theoretical oscillator models illustrated by symbolic schemes. Our aim here is two-fold: (1) To provide a qualitative description of the essential elements underlying most biological oscillators; and (2) examine in detail an endocrine case study of the growth hormone network in the male rat.

Essential Elements for Oscillatory Behavior. For the purposes of this laboratory project, we can define oscillation as a pattern in the dynamic plot of a measurable quantity [the variable under consideration (e.g., population size, concentration of a particular hormone in the blood stream, and so forth)] that exhibits a recurring waveform. An important characteristic of an oscillation is its period (or interpulse interval), which is simply the interval between two identical reference points on the pattern (Figure 11-1, left panel). If the amplitude range of the oscillation is continuously decreasing, we have a case of damped oscillation (Figure 11-1, right panel). The repeating pattern is not limited to simple sine- or cosine-like waveforms and could be more complex (see Figure 11-5). If an oscillation system contains more than one oscillation variable, a critical characteristic is the phase relationship between the variables.

For most biological systems to exhibit rhythmic behavior, one or both of the following qualitative behaviors need to occur:

---

1. From Friesen, W. O., & Block, G. D. (1984). What is a biological oscillator? *AJP: Regulatory, Integrative, and Comparative Physiology 246*, R847–R853. Used by permission.

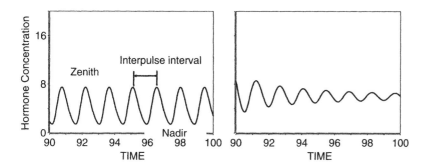

FIGURE 11-1.
Left panel: oscillations with constant amplitude; right panel: damped oscillations.

1. Delay

Recall that if a system has a stable equilibrium state, its values stabilize around this state in the long run (i.e., the limit as $t \to \infty$ of the variable as a function of time is equal to the equilibrium state). An oscillatory system must include a restorative process that, like inertia in physical systems, leads to overshoot of the equilibrium value. For most cellular, endocrine, and neuronal oscillations, the factor that provides an overshoot is *delay*. The element of delay prevents the restorative process from coming into full play until the equilibrium value has been passed. This situation is illustrated in Example 11-1.

Delays in biological systems can arise from many sources, and the debate about what causes delay and how best to model complex systems involving delay is ongoing. By far, the simplest situations involving delay are those in which there is a certain time-offset or lag between an action triggered by one variable and the response to this action by a second variable. For example, suppose hormones A and B are involved in the control of a particular organismal function, and that hormone B turns off the production of hormone A. Let us also suppose that the inhibitory effect of hormone B does not immediately follow after an increase of hormone B in the bloodstream. The time elapsed between the increase in the concentration of B and the decrease in the concentration of A will represent an *explicit delay*. An explicit delay would thus represent the amount of time necessary for a certain sequence of molecular and/or cellular events to occur. Explicit delays can vary greatly, depending upon the system at hand—the incubation period for an infectious disease represents one kind of explicit delay, and incubation periods can range from days to years. In this case, the delay results from the amount of time it takes for the pathogen to travel through the host's body and to multiply in the favored portion of the host's anatomy.

In other cases, the delay may not be explicit. In such a case, the delay would not reflect a certain period of time but would result from a particular threshold value that must be met before the affected variable responds. In our hormone example, let us say that the level of hormone A will not be affected until the level of hormone B in the bloodstream reaches 300 pg/ml (picograms per milliliter). For another example, consider the Lotka–Volterra predator–prey model. This represents a coupled system with threshold values for the predator and prey populations. The population of the predator (the owl) exerts control over the population of the prey (voles) and vice versa. In our work with this model in Chapter 2 of the text, the thresholds were manifested as critical lines that were determined by the rate of change of the variables being equal to zero (null clines).

2. Negative feedback

If a system involves more than one variable and the variables interact with one another, it is possible that an increase in one of the variables may inhibit its own growth by interacting with the other variable. This is, for example, the case in predator–prey models of the Lotka–Volterra type (see Example 10-2). Such interaction is referred to as *negative feedback* and is often a factor that creates oscillations. When a single variable is considered, a self-inhibitory feedback is sometimes possible—in this case, we talk about *negative autofeedback* (see Example 11-1).

It is imperative, however, to understand that whether delay and/or negative feedback will trigger oscillatory behavior depends on the specific context of the problem and on the particular values of the system's parameters. In this light, the presence of delay and/or negative feedback should be considered a factor that is *likely* to cause oscillation and should by no means be understood as a sufficient condition for oscillatory behavior in biological systems.

Symbolic-Scheme Representations of Theoretical Models. Schematic diagrams are often used to show explicitly the most important components of a biological system. The following symbols have been adopted in order to facilitate the display of information: (1) Rectangles (A, B, etc.) are used to denote the system's variables; (2) lines indicate causality and are additionally marked with one of the symbols (+) or (−) [a (+) indicates an excitatory action on the variable at which the line terminates, whereas a (−) denotes an inhibitory action]; and (3) triangles (D) on one or more of the lines indicate that some delay occurs from the rise in the variable from which the line initiates until the effect indicated by the line is actually exerted. For example, Figure 11-2 presents a schematic diagram of a network representing the interaction between two variables A and B. The line from A to B is marked with a (+), indicating an excitatory input. The line from B to A is marked with a (−) to indicate that B inhibits the growth of A (negative feedback). In addition, there is the presence of explicit delay for the action of A upon B.

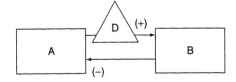

FIGURE 11-2.
Schematic diagram of a two-node network with feedback and delay.

EXAMPLE 11-1

Delayed Population Growth. The classical logistic growth equation is

$$\frac{dP}{dt} = r\left(1 - \frac{P(t)}{K}\right)P(t),$$

where $K > 0$ is the carrying capacity of the system, $r > 0$ is the maximal per capita rate of growth for the population, and $P(t)$ is the size of the population at time t. This equation presents a model in which the population size is limited by the available resources. A major disadvantage of this model is that it fails to take into consideration the time necessary for complex organisms to reach reproductive age. (Why?)

The diagram in Figure 11-3 illustrates a self-inhibitory effect (autofeedback) for the population size delayed by time D necessary for each individual to reach reproductive age.

Notice that the origin of the line indicating excitatory input for the population size $P(t)$ is not shown. The input here is generated by the flow of natural resources that support the living organisms in the system.

A differential equation with delay D related to the classical logistic growth equation (sometimes called Hutchinson's equation) is:

$$\frac{dP}{dt} = r\left(1 - \frac{P(t - D)}{K}\right)P(t).$$

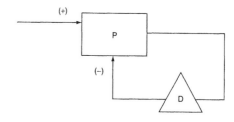

FIGURE 11-3.
Schematic representation of autofeedback with delay.

EXAMPLE 11-2

A Predator–Prey Model. The diagram in Figure 11-4 represents schematically the Lotka–Volterra predator–prey model. Because the owls feed on the voles, the growth of the vole population causes growth in the owl population (excitatory input). Because the growth of the owl population causes the vole population to decline in size (inhibitory input), there is a negative feedback in this system.

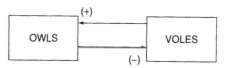

FIGURE 11-4.
Schematic presentation of the feedback in the Lotka–Volterra predator–prey model.

B. Modeling of Hormone Networks

There is overwhelming evidence that the pattern of hormone delivery to target organs is crucial to the effectiveness of their action. If the required hormone is supposed to be delivered in a pulsatile pattern and is instead delivered at a constant level, the target organ will likely not respond adequately. If the required hormone is supposed to be delivered in 24 pulses per day and instead is delivered in 12 pulses, the target organ probably will not respond appropriately. In normal, healthy individuals, differences in hormone release patterns have been observed between adult males and females and between juveniles and adults. Hormone release patterns may be altered because of disease, and the consequent endocrine changes may result in significant morphological and physiological changes. Thus, it should come as no surprise that the mechanisms controlling hormone dynamics are increasingly the subject of biomedical research.

Reaching an understanding of endocrine systems is made particularly difficult by their high complexity. There may be multiple interactions, both positive and negative, both with and without delays. Consequently, quantitative methods have been developed to complement qualitative analysis and laboratory experiments with the goal of revealing the specifics of hormone release control. The emerging mathematical models interpret endocrine networks as dynamic systems and attempt to simulate and explain their temporal behavior. The sources Farhy et al. (2002), Farhy and Veldhuis (2004), Keenan and Veldhuis (2001a) and (2001b); Chen et al. (1995), and Wagner et al. (1998) can be used as a starting point. For example, it is well known that the concentration of many hormones in the bloodstream rises and falls multiple times per day and exhibits a pulsatile pattern. Thus, the mathematical models describing the dynamics of these hormones should be able to capture and explain the oscillatory behavior they exhibit.

In this project, we focus on the endocrine network of the growth hormone (GH). GH is secreted by the pituitary gland under the control of the following substances released by the hypothalamus: (1) the GH-releasing hormone (GHRH) that triggers the production and secretion of GH, and (2) somatostatin, or somatotropin release-inhibiting factor (SRIF), that is a GH release inhibitor. Numerous other substances could impact GH behavior. However, they are considered to be of secondary importance to GH network and, for the sake of simplicity, will be omitted in the discussion that follows.

C. A Case Study: The GH Network and the GH Secretion Pattern in Male Rats

Frequent serial measurements of peripheral GH concentrations have unmasked complex patterns of gender-specific and developmentally regulated GH release in rats, sheep, and humans. A thorough list of references can be found in Farhy and Veldhuis (2004). In particular, GH secretion evolves as infrequent clusters of high pulses in the adult male rodent, pubertal children, and young fasting or sleeping men and women. However, GH secretion consists of frequent, isolated, low-amplitude

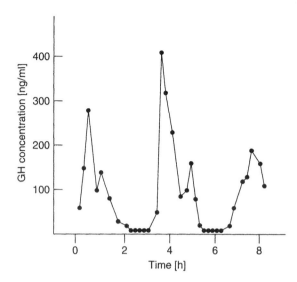

FIGURE 11-5.
A typical GH secretion profile in the adult male rat. For actual in vivo data see, for example, Lanzi and Tannenbaum (1992).

bursts in the female rat and older, awake, or nutrient-replete humans. Figure 11-5 presents a typical GH secretion profile in the adult male rat.

In a simplified view, GH secretion increases with increased hypothalamic GHRH secretion but is inhibited by SRIF (which acts as a suppressor for both GHRH and GH). All three hormones are subject to exponential elimination with known half-lives. Laboratory data in the male rat show that elevated concentrations of GH act by way of time-delayed feedback ($D = 40$ to 60 minutes) to stimulate SRIF secretion from the hypothalamus into the bloodstream. This, in turn, serves to antagonize GH release by the anterior pituitary gland, leading to the widely accepted Consensus GH Control Network (Plotsky and Vale [1985]; Robinson [1991]) depicted in Figure 11-6.

Network Characteristics. Laboratory experiments suggest the following: (1) GH/SRIF feedback drives low-frequency composite GH volleys that recur every 3 to 3.3 hours; (2) there is an underlying high-frequency oscillatory GH rhythm corresponding to period of 45 to 60 minutes; (3) the GH concentration remains below 1000 pg/ml at all times (normal concentrations are usually below 500 pg/ml but concentrations as high as 1000 pg/ml are possible under special circumstances for short periods of time); (4) GHRH remains below 1000 pg/ml; (5) SRIF varies between 0 and 150 pg/ml; and (6) the elimination half-lives of GH, GHRH, and SRIF are ln2/2.7 hours, ln2/8 hours, and ln2/5 hours, respectively.

Identifying a Problem. Regardless of its wide popularity, when implemented as a mathematical model, the Consensus GH Control Network fails to explain the high-frequency oscillatory component in the GH release profile. Consequently, regardless of the specific parameters of the system, the

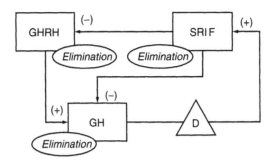

FIGURE 11-6.
Consensus GH network.

Consensus GH Control Network fails to explain the specific GH dynamic pattern presented in Figure 11-5. Thus, the following question arises:

Question: Could the discrepancy be resolved within the scheme of the Consensus GH Network?

The following putative link has been considered by Farhy and Veldhuis (2004) in search of an answer:

Conjecture: GHRH stimulates SRIF secretion with some delay to account for the high-frequency pulses.

In the exercises that follow, you will be asked to examine this conjecture. We begin with a review of the mathematical background.

MATHEMATICAL BACKGROUND

To describe the quantitative approximation of the concentration dynamics of a single hormone, we assume that the hormone concentration rate of change depends on two processes—secretion and ongoing elimination. The quantitative description is given by the differential equation

$$\frac{dC}{dt} = -kC(t) + aS(t), \tag{11-1}$$

where $C(t)$ is the hormone concentration, t is the time, k is the rate elimination constant of the hormone [related to the half-life by *half-life* $= \ln(2)/k$], and $S(t)$ is a function with values between $[0,1]$ that models the rate of secretion. The parameter a is related to the maximal attainable concentration value in the following way: $a = kC_{max}$ (see Exercise 11-2). As is evident from the model, elimination is assumed to be proportional to the concentration.

SOFTWARE BACKGROUND

The software necessary for this lab is *BERKELEY MADONNA*. The Fast Fourier Transform (FFT) feature will be helpful for identifying the periodic components of the oscillations. It transforms the data into frequency domain and allows determination of the contribution to the data of each periodic component. The peaks in the frequency graph determine the frequencies of the periodic components, with higher peaks corresponding to the more prevalent frequencies. More details regarding the Fourier transform of a time series can be found in Chapter 9 of the text.

To view the Fourier transform of your data, activate the FFT button that can be found across from the Run button in the Graphics window and features an icon with the letter F. The graph of the data in the frequency domain will then be displayed. You can "zoom in" to take a better look at a portion of the data. This can be done by clicking and dragging the mouse to form a rectangle over the portion of the data you wish to view. Releasing the mouse will then cause the selected area to expand to fill the entire Graphics window. To zoom back out, use the Zoom-out button featuring an icon with the letter Z.

To find the period and frequency, activate readout mode by clicking the Readout button that is positioned across from the Run button, featuring an icon representing a cross. Place the mouse pointer over the lowest frequency peak, and press the mouse button. The frequency, period, and amplitude will be displayed in the upper-right corner of the window.

PART I. SIMULATING THE CONCENTRATION DYNAMICS OF A SINGLE HORMONE THAT DEPENDS ON ANOTHER HORMONE

If we first assume that that the secretion rate $S(t) = S_A$ of hormone A does not depend explicitly on t but is, instead, controlled only by the concentration of some other hormone B, then S from Eq. (11-1) will be a function of the concentration $C_B(t)$ of the hormone B. We write $S_A = S_A(C_B(t))$. From now on, the function S will be called a *control function*. The choice of the control function, albeit arbitrary to some extent, should conform to a set of general rules. For example, the control function must be non-negative since the secretion rate is always non-negative. Also, in most cases, the secretion function will be monotone. A monotone increasing control function represents positive interaction, and a monotone decreasing control function represents a negative interaction.

Many authors use the following nonlinear, sigmoid functions, known as *up- and down- regulatory Hill functions*, to describe positive and negative interaction, respectively:

$$F_{up}(C) = \frac{C^n}{C^n + T^n}; \quad F_{down}(C) = \frac{T^n}{C^n + T^n}.$$

The parameter $T > 0$ is called a threshold, and the exponent $n \geq 1$ is called a Hill coefficient. Both functions are monotone and map the interval $(0,\infty)$ to $(0,1)$. The threshold T in an up-regulatory Hill function is sometimes denoted by ED_{50} and is called the *median effective dose*. Analogously, in a down-regulatory control function, T is referred to as the ID_{50}, or *median inhibitory dose*. The parameters ED_{50} and ID_{50} approximate the potency of the regulatory hormone. The Hill coefficient n controls the slope. The larger the value of n, the steeper the slope, and, for large n (values as large as $n = 100$ exist in biology), the control function acts almost as an on/off switch at the concentration value $C = T$. The Hill functions F_{up} and F_{down} are exemplified in Figure 11-7 for $n = 3$ and $T = 10$.

EXAMPLE 11-3

Write a system of differential equations describing the schematic network in Figure 11-2 if A and B are hormones eliminated in the bloodstream.

SOLUTION:

To simplify the notation, let $A = A(t)$ and $B = B(t)$ denote the concentrations of the hormones A and B, respectively [i.e., we are using $A(t)$ instead of $C_A(t)$ and $B(t)$ instead of $C_B(t)$]:

$$\frac{dA}{dt} = -k_1 A + a_1 \frac{T_1^{n_1}}{B(t)^{n_1} + T_1^{n_1}}$$

$$\frac{dB}{dt} = -k_2 B + a_2 \frac{(A(t-D))^{n_2}}{(A(t-D))^{n_2} + T_2^{n_2}}.$$

We have used a down-regulatory Hill function with parameters T_1 and n_1 in the first equation, since the concentration of A is inhibited by B. We

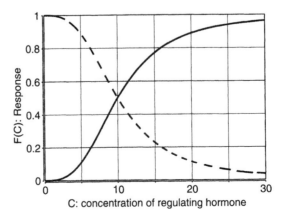

FIGURE 11-7.
Exemplary profiles of up-regulatory (black line) and down-regulatory (dashed line) Hill functions. In both examples, $n = 3$ and $T = 10$.

have used an up-regulatory Hill function with parameters T_2 and n_2, and delay D for the second equation, because the concentration of B is stimulated by the concentration of A with a delay of D time units.

Exercise 11-1
....

Show that $F_{up} = 1 - F_{down}, F_{up}(T) = 1/2,$ and $F_{down}(T) = 1/2.$

Exercise 11-2
....

Show that if in Eq. (11-1) the control function S is a Hill function, the concentration function $C(t)$ cannot exceed the value $C_{max} = a/k$. This shows that the control coefficient a in Eq. (11-1) is determined as the product of the maximal hormone concentration and the elimination rate constant; that is, $a = kC_{max}$.

Hint: Use that the minimal and maximal concentrations are achieved when $dC/dt = 0$ in Eq. (11-1). Combine this with the fact that the values of a Hill function never exceed 1.

Exercise 11-3
....

Write the differential equations describing the (putative) two-hormone network depicted in Figure 11-8. When sufficient information exists about some of the model parameters, determine their numerical values. For example, the numerical values for the elimination constants, as well as the coefficients determining the maximal hormone concentration, should be specified. In your solution, identify these values clearly.

Hint: Follow the steps outlined below:

(a) Denote the concentration of GHRH by $GHRH(t)$.

(b) Denote the concentration of SRIF by $SRIF(t)$.

(c) For the concentration of GHRH, use the following equation

$$\frac{d(GHRH)}{dt} = -kGHRH + aS(SRIF),$$

where a controls the maximal attainable amplitude of $GHRH$ and is measured in concentration/time, S is the secretion control function, k

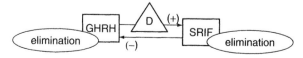

FIGURE 11-8.
A two-hormone network.

is the elimination rate constant [related to the half-life by the formula $k = \ln 2 /(\text{half-life})$].

(d) Use the information under Network Characteristics from Section C above to determine the value of the elimination rate constant k.

(e) Use Exercise 11-2, the half-lives determined in (d), and the maximal concentration values for GHRH and SRIF (as specified under Network Characteristics from Section C above) to determine the values of the control function coefficients.

(f) Use a Hill function (as in Example 11-3) for the control function S.

(g) Follow a similar sequence of steps to implement the differential equation that describes the rate of change for the concentration of SRIF.

(h) Implement the delay with a constant value of D hours.

Exercise 11-4

Enter your model from Exercise 11-4, into *BERKELEY MADONNA* using a starting value of 14 minutes (0.23 hours) for D. For the control functions for GHRH and SRIF, use initial threshold values of $T_1 = 20$ pg/ml and $T_2 = 780$ pg/ml, respectively. Use sliders to adjust the system parameters (the Hill coefficients, the thresholds in the control functions, and the delay D) in a way that causes oscillatory behavior for the GHRH with pulses recurring every 60 minutes. For the Hill functions, consider values less than $n = 5$.

Hint: Using the FFT plot of the data as described in the Software Background section of the project will be helpful here. Activate the FFT button and use the frequency plot of the solution to adjust the system parameters in a way that will generate periodic behavior with a period of about 60 minutes.

Exercise 11-5

Preserving all parameter values from Exercise 11-4, incorporate the GH–GHRH conduit that appears in Figure 11-9 to get GH pulses every 60 minutes. In your report, present (1) the system of differential equations; (2) the *BERKELEY MADONNA* code and program output; and (3) the specific values for the parameters that correspond to the state of the system that gave GH pulses every 60 minutes. For the GH control function, use a starting value of $T_3 = 780$ pg/ml and Hill coefficients less than 5.

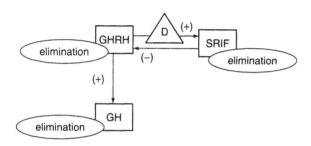

FIGURE 11-9.
Another GH network.

PART II. SIMULATING THE CONCENTRATION DYNAMICS OF A SINGLE HORMONE THAT DEPENDS ON MORE THAN ONE OTHER HORMONE

Let's assume now that two hormones control the secretion of A instead of one. We denote them by B and C with corresponding concentration functions C_B and C_C. The control function S_A is now a function of C_B and C_C [i.e., $S_A = S_A(C_B, C_C)$ depending on the specific interaction between A from one side, and B and C from another]. We list a few special cases here.

For example, if both B and C stimulate the secretion of A,

$$S_A(C_B, C_C) = k_1 F_{up}(C_B) + k_2 F_{up}(C_C),\qquad(11\text{-}2)$$

when B and C act independently, or

$$S_A(C_B, C_C) = F_{up}(C_B) F_{up}(C_C),\qquad(11\text{-}3)$$

when B and C act simultaneously (that is, when the secretion of A requires the presence of both). In Eq. (11-2), the parameters k_1 and k_2 are such that $k_1 + k_2 = 1$. These parameters are weights that reflect the degree to which hormones B and C each contribute to the control of hormone A. In case they contribute equally, $k_1 = k_2 = \frac{1}{2}$. On the other hand, if the secretion of A is stimulated by B but is suppressed by C, and B and C control simultaneously the secretion of A, the control function could be introduced as

$$S_A(C_B, C_C) = F_{up}(C_B) F_{down}(C_C).\qquad(11\text{-}4)$$

Exercise 11-6

Preserving all parameter values from Exercise 11-5, add two conduits, namely, the SRIF–GH negative interaction and the delayed GH → SRIF link as shown on the diagram in Figure 11-10. This is the Consensus GH Network from Figure 11-6 with the added putative link between GHRH

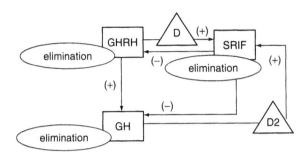

FIGURE 11-10.
Another (more complex) GH network.

and SRIF, as conjectured in Section C above. Recall that D_2 is approximately 60 minutes.

- You may assume that GHRH and SRIF affect the GH concentration simultaneously [i.e., assume that the control function for the GH is a product of two Hill functions, as in Eq. (11-3)]. Use a threshold value of $T_5 = 20$ pg/ml for the control function corresponding to the SRIF $\rightarrow$ GH link.

- Assume that GH and GHRH affect the concentration of SRIF independently. Thus, the control function in this case will be a sum of two Hill functions, as in Eq. (11-2). Use a threshold value of $T_4 = 110$ pg/ml for the control function corresponding to the GH $\rightarrow$ SRIF link.

Adjust the system parameters to get a profile that resembles the one shown in Figure 11-5. In your report, present (1) the system of differential equations, (2) the *BERKELEY MADONNA* code and program output, and (3) the specific values for the parameters corresponding to the output.

Exercise 11-7

Modify your solution for Exercise 11-6 by *removing* the conjectured GHRH $\rightarrow$ SRIF delayed link, thus obtaining the equations describing exactly the Consensus GH Network as shown in Figure 11-6. Try to adjust the system parameters, and attempt to obtain the GH concentration profile from Figure 11-5. Were you successful?

REFERENCES

Chen, L., Veldhuis, J. D., Johnson, M. L., & Straume, M. (1995). Systems-level analysis of physiological regulatory interactions controlling complex secretory dynamics of the growth hormone axis: A dynamic network model. *Methods in Neurosciences* (vol. 28, pp. 270–310). New York: Academic Press.
Farhy, L. S., Straume, M., Johnson, M. L., Kovatchev, B. P., & Veldhuis, J. D. (2002). Unequal autonegative feedback by GH models the sexual dimorphism

in GH secretory dynamics. *American Journal of Physiology Regulatory, Integrative and Comparative Physiology, 282,* R753–R764.

Farhy, L. S., & Veldhuis, J. D. (2004). Putative growth hormone (GH) pulse renewal: Periventricular somatostatinergic control of an arcuate-nuclear somatostatin and GH-releasing hormone oscillator. *American Journal of Physiology Regulatory, Integrative and Comparative Physiology, 286,* R1030–R1042.

Friesen, W. O., & Block, G. D. (1984). What is a biological oscillator? *American Journal of Physiology, 246,* R847–R853.

Keenan, D. M., & Veldhuis, J. D. (2001). Hypothesis testing of the aging male gonadal axis via a biomathematical construct. *American Journal of Physiology Regulatory, Integrative and Comparative Physiology, 280,* R1755.

Keenan, D. M., & Veldhuis, J. D. (2001). Disruption of the hypothalamic luteinizing hormone pulsing mechanism in aging men. *American Journal of Physiology Regulatory, Integrative and Comparative Physiology, 281,* R1917.

Lanzi, R., & Tannenbaum, G. (1992). Time course and mechanism of growth hormone's negative feedback effect on its own spontaneous release. *Endocrinology, 130,* 780–788.

Plotsky, P. M., & Vale, W. W. (1985). Patterns of growth hormone-releasing factor and somatostatin secretion into the hypophysial-portal circulation of the rat. *Science, 230,* 461–463.

Robinson, I. C. A. F. (1991). The growth hormone secretory pattern: a response to neuro-endocrine signals. *Acta Paediatrica Scandinavica. Supplement, 372,* 70–78.

Wagner, C., Caplan, S. R., & Tannenbaum, G. S. (1998). Genesis of the ultradian rhythm of GH secretion: A new model unifying experimental observations in rats. *American Journal of Physiology, 275,* E1046–E1054.

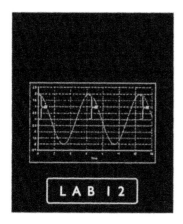

Chemical Perturbation on the Operations of Circadian Clocks

The experimental data collected and used to analyze circadian rhythms are often *confounded*. What this means is that extraneous issues (i.e., effects distinct from those of known experimental variables) are responsible for altering, in an uncontrollable and unknown manner, the appearance and properties of experimentally observed data (go ahead and look up the dictionary definition of the word "confound"). A useful analogy may be the following: In performing experiments in biological research, we seek to "see" the biology underlying the hypothesis for which answers are being sought. However, our "views" of this underlying biology are obscured, or confounded, by "dirty windows" through which we must observe this biology. The dirty windows arise as a consequence of our imperfect ability to experimentally monitor the biological phenomena we seek to measure. Therein lies our challenge.

Confounded data may contain (1) mean and/or variance nonstationarities (i.e., time-dependent drifting and/or changes in oscillatory amplitude); (2) period and/or phase instability; and/or (3) noise. In such cases, analyzing the rhythmic nature of the data may be challenging. For this laboratory project, you will use simulated data meant to represent *per-luc* activity in an in vitro sample before and after treatment with a translational inhibitor. The goal is to determine the effect of the translational inhibitor on the circadian behavior of the cell.

In this lab, you will:

- Learn to recognize mean and variance nonstationarities in data.
- Gain working knowledge on interpreting the outputs from various rhythm analysis algorithms such as *ARFILTER, DTRNDANL, PHASEREF,* and *FFT-NLLS*.
- Use those algorithms to obtain estimates of the period, amplitude, and phase of the rhythmic data.
- Use those algorithms to estimate the robustness of rhythmic expression (i.e., to determine how strongly rhythmic is the observed patterning) or, conversely, to conclude that the series is arrhythmic.
- Investigate how a transcriptional inhibitor would affect the circadian clock mechanisms of the cells.
- Experience independent, research-like exploration of an open-ended problem.

BIOLOGICAL BACKGROUND

The molecular mechanisms controlling circadian rhythms involve the expression of several known genes. As is shown in Figure 12-1, in the *Drosophila* circadian clock, the genes *tim*, *per*, *dClk*, *cyc*, and *dbt* are involved in the mechanism. Gene expression requires both transcription (RNA synthesis) and translation (protein synthesis). The *tim* and *per* genes result in the production of TIM and PER proteins, the *dClk* and *cyc* genes produce the dCLK and CYC proteins, and the *dbt* gene produces the DBT protein. TIM and PER form a complex that travels to the nucleus and interferes with the transcription of the *dClk* and *cyc* genes. Light causes the degradation of the TIM/PER complex and, when the complex is gone, the *cDlk* and *cyc* genes are transcribed and their proteins, dCLK and CYC, are produced. These proteins also bind to each other and form a complex that travels to the nucleus. dCLK and CYC are transcription factors, and they turn on the expression of the *per* and *tim* genes. The last protein shown on the diagram is the *dbt* gene product DBT, which is a protein kinase, an enzyme which phosphorylates proteins. DBT phosphorylates the free (uncomplexed) PER protein, precipitating its destruction.

Assessment of the regulatory patterning of the *period* circadian clock gene (*per*) became possible through the use of a reporter gene—in this case, a luciferase cDNA fused to the promoter region of *per* (see Figure 12-2). Luciferase is the enzyme responsible for bioluminescence—when luciferase acts upon its substrate, luciferin, light is produced. This recombinant reporter gene DNA, called a *per-luc* construct, was introduced into rat zygotes to produce what are called transgenic rats, which have the *per-luc* DNA in every cell of their bodies. These rats are

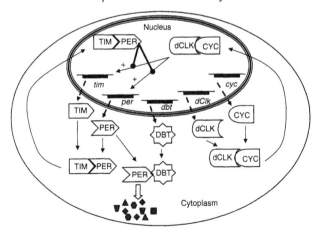

FIGURE 12-1.
Simplified schematic diagram of the *Drosophila* circadian clock mechanism. Abbreviations are as follows: *tim* and TIM are the *timeless* gene and protein, respectively; *per* and PER stand for *period* gene and protein, *dbt* and DBT for *double time*; *dClk* and dCLK for *clock*; *cyc* and CYC for *cycle*. The *doubletime* protein DBT phosphorylates free PER protein (i.e., any PER not bound to TIM) and facilitates its degradation. The dashed lines represent transcription followed by translation. In the nucleus, the arrows indicated by a (+) indicate the elevation of transcription and the arrows with the round heads indicate blocking of this activity.

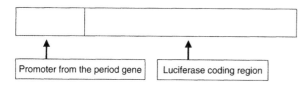

FIGURE 12-2.

Schematic diagram of a *per-luc* construct. The DNA from the controlling region of the *period* gene (the *period* promoter) has been fused to the DNA making up the coding region of the luciferase gene, producing a luciferase which is then under circadian control.

entrained to a circadian rhythm, and then the tissue to be studied is excised from the animal and put into tissue culture. Light emission from the cultured tissue is then observed.

In this laboratory project, we are going to ask what would happen to the circadian rhythms in the cultured cells if they were treated with a translational inhibitor. A *translational inhibitor* is an agent that interferes with the functioning of the ribosomes, the protein synthetic machinery. As long as it is present in the cells, there will be no translation. When it is removed from the cells, translation can begin again. You will be presented with data sets from before and after treatment with a translational inhibitor, and your task will be to determine the effects on the circadian behavior of the cells.

MATHEMATICAL BACKGROUND

A. Confounded Data

One challenge presented by bioluminescence time series data generated from *per-luc* experiments has to do with depletion with time of luciferin, the substrate necessary for light production by the luciferase enzyme. Consequently, time-series traces of experimental data show patterning in which average bioluminescence intensity is often decreasing with time (drifting downward). Figure 12-3 illustrates this situation. Also, in cases when the method of bioluminescence has been applied to studies of circadian phenomena, a situation is often encountered in which the oscillatory magnitude changes with time. These two examples illustrate various cases of nonstationarities. In general, a time series is *stationary* if its mean, variance, and autocorrelation remain constant through time. Figure 12-3 represents a nonstationary time series, because its mean and variance change with time.

Sometimes, as in the work of Plautz *et al.* (1997), a secondary reproducible peak in the data can be observed, suggesting either an approximately half-circadian period component and/or the presence of two circadian rhythms out of phase with each other by approximately half a circadian period. Analysis by any method that assumes a single period estimate near a circadian range (i.e., in the vicinity of 24 hours) would be unable to accurately capture the dynamic behavior exhibited by this type of occurrence.

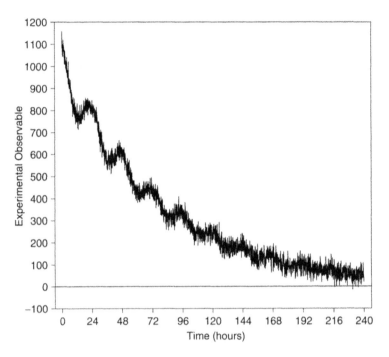

FIGURE 12-3.
A time series exhibiting a trend and variance nonstationarity.

Finally, additional problems in the rhythm analysis of time series come from the presence of noise in the data that may often obscure the circadian rhythms. As we shall see later, this is particularly pertinent to assessing the phase information of a rhythm from a noise-confounded time series. In such cases, algorithms that would allow using appropriate filters in order to "clean" the data may be helpful.

B. Fundamentals of Rhythmic Data

Recall that a purely periodic waveform might be of the form:

$$f(t) = \alpha \cos(\beta t + \gamma).$$

Each of the parameters α, β, and γ serves a particular function. The parameter α controls the amplitude of the wave [see Figure 12-4 (A), which shows the graph of $\alpha\cos(t)$ where $\alpha > 0$]. The parameter β controls the period (and therefore the frequency, because the frequency is the inverse of the period). In Figure 12-4(B), we give the graph of $\cos(\beta t)$. Note that the larger β is, the higher the frequency and the shorter the period.

The parameter γ is the phase shift of the graph. It shows how much the graph is shifted to the left ($\gamma > 0$; an advance) or the right ($\gamma < 0$; a delay). It can also be interpreted as the time value at which the periodic or rhythmic pattern is at a maximum (when modeling in terms of a cosine wave). The graph of $\cos(t + \pi/4)$ is shown in Figure 12-5(A), and the graph of $\cos(2t + \pi/4) = \cos(2(t + \pi/8))$ is shown in Figure 12-5(B).

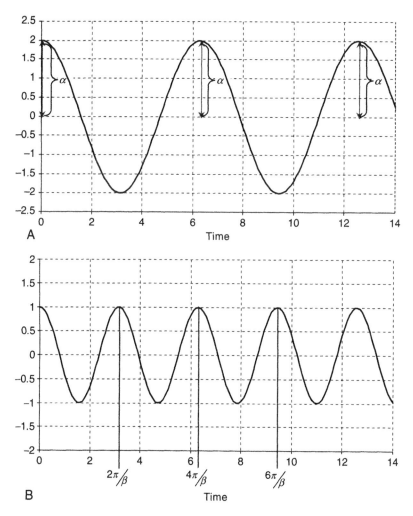

FIGURE 12-4.
Graphs of $\alpha\cos(t)$ with $\alpha > 0$ (panel A) and $\cos(\beta t)$ with $\beta > 1$ (panel B).

A cosine-like pattern (or possibly a truncated such pattern) is often found in circadian rhythm data, but there are also some variations. The amplitude might vary from one cycle to another such as in the graph shown in Figure 12-6(A), and the period may change by compressing (or expanding) such as in the graph shown in Figure 12-6(B).

One characteristic often critical in circadian biology research regards assessing phase shifts of expression patterning in response to perturbing stimuli of various natures. Such assessments require an assumption of a constant periodicity in the rhythmic patterning being considered (i.e., approximately 24 hours for circadian rhythms). Consider the graph shown in Figure 12-7(A), which is periodic with an exact period of 24 hours. In Figure 12-7(B), we have plotted the phase optimum for each cycle. In the context of circadian rhythms, this can be interpreted as the time of day at which the maximum value (or acrophase) occurs. If there is no change in the period, the maximum value will always occur at the same time every day and the plot will be composed of values arranged

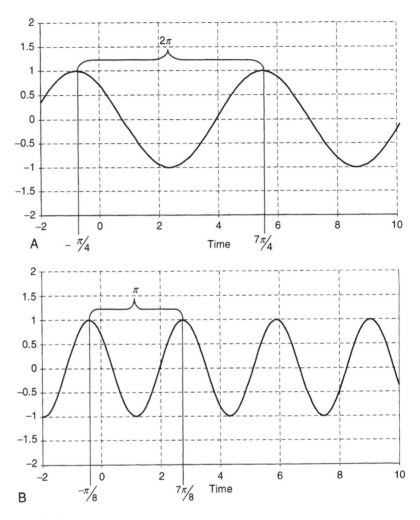

FIGURE 12-5.
Graphs of $\cos(t + \pi/4)$ (panel A) and $\cos(2t + \pi/4) = \cos(2(t + \pi/8))$ (panel B).

in a horizontal line [see Figure 12-7(B)]. Now consider the graph in Figure 12-7(C), where the phase is delaying by a constant amount each day. Figure 12-7(D) shows the time at which the maximal value occurs for each day. As the data points rise, this means that the maximal value occurs later and later each day. In this particular example, the data also indicate that the phase delays in a linear fashion, by about 2.6 hours per day. In the exercises which follow, we shall observe other functional dependencies.

SOFTWARE BACKGROUND

You will be working with the output from rhythm analysis software developed in the Center for Biomathematical Technology at the University of Virginia. We present a brief description of the procedures here, but our focus for this project will be on interpreting the results obtained by these procedures.

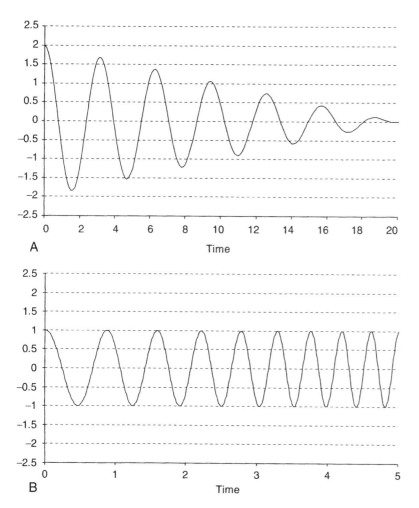

FIGURE 12-6.
Cosine-like rhythmic pattern with decreasing amplitude (panel A) and shrinking period (panel B).

FFT-NLLS. This algorithm analyzes data for the presence of multiple periodic components in the form of a sum of multiple cosine waves. It combines two fundamental techniques that have been discussed previously—FFT methods and NLLS methods. The *FFT-NLLS* procedure can be outlined as follows: time series are first linear regression detrended by *FFT-NLLS* to produce zero-mean, zero-slope data. An FFT power spectrum is then calculated of the detrended data. The period, phase, and amplitude of the most powerful spectral peak are used to initialize a one-component cosine function (i.e., $n = 1$ is used, initially) of the form:

$$y_{LR}(t) = c \sum_{i=1}^{n} + \alpha_i \cos\left[\frac{2\pi(t + \phi_i)}{\tau_i}\right]$$

in which $y_{LR}(t)$ is the linear regression detrended time series to which analysis is being performed, c is a constant offset term, n is the order of fit, t is time, and α_i, φ_i, and τ_i are the amplitude, phase, and period,

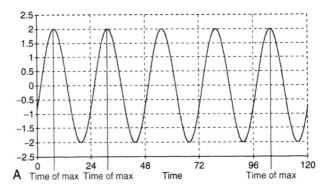

A

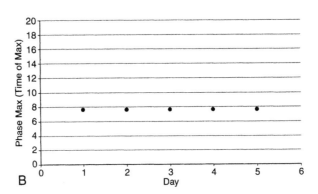

B

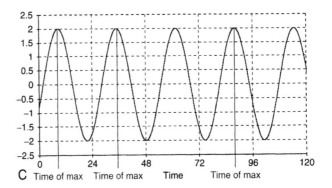

C

FIGURE 12-7.
For the graph in panel A, the period does not change. In panel B, the acrophase (the point of daily maximum) occurs at the same time each day. In panel C, the period is expanding. In panel D, the daily maximum is delayed by about 2.6 hours from day to day.

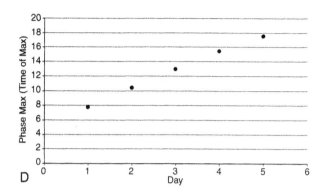

FIGURE 12-7. Cont'd.

respectively, of the *i*-th cosine component. The parameters of this
function are then estimated by nonlinear least-squares minimization.
Upon convergence, approximate nonlinear asymmetric joint confidence
limits are estimated for all parameters (period, phase, amplitude,
and constant offset) at a 95% confidence probability. If the amplitude is
significantly different from zero, then the procedure is repeated at the
next higher order. The two most powerful FFT spectral peaks are then
used to initialize a two-component cosine function (i.e., $n = 2$), which is
subsequently NLLS-minimized to the linear regression detrended data,
and confidence limits are again evaluated. This process is repeated
iteratively until at least one cosine component is identified with an
amplitude that is not statistically significant.

The statistical significance of each derived rhythmic component is
assessed by way of the relative amplitude error (RAE). The RAE is
defined as the ratio of, in the numerator, the amplitude error (one-half
the difference between the upper minus the lower 95% amplitude
confidence limits) to, in the denominator, the most probable derived
amplitude magnitude. Theoretically, this metric will range from 0.0 to
1.0; 0.0 indicates a rhythmic component known to infinite precision (i.e.,
zero error), 1.0 (or greater) indicates a rhythm that is not statistically
significant [i.e., error equal to (or exceeding) the most probable
amplitude magnitude], and intermediate values indicate varying
degrees of rhythmic determination. Thus, the lower the RAE value, the
more well-determined the respective rhythmic component is.

As an example, consider the output in Figure 12-8 from *FFT-NLLS* on
time-series data from a file named *TRNSPLNT.NO*. This data file can be
downloaded from www.biomath.sbc.edu/data.html. For now, we shall
only focus on the first four columns of output (and will only use some of
the information that appears there).

The first three columns present the amplitude, period, and phase
estimates for each of the periodic components identified by the
algorithm. For the purposes of this project, the estimates for the period and
phase will be the most important. The fourth column presents the RAE that
the algorithm assigns to each periodic component. As discussed above, the

FFT-NLLS of TRNSPLNT.NO

File Name	Amplitude	Period	Phase	Rel-Amp	Rel-Per	Rel-Phi
trnsplnt.no	3.574E+01	2.000E+01	7.950E+00	0.078	0.001	0.014
	(2.773E+00)	(2.808E-02)	(2.726E-01)			
	1.316E+01	2.453E+01	-1.147E+01	0.212	0.006	0.048
	(2.789E+00)	(1.455E-01)	(1.169E+00)			
	1.026E+01	9.982E+00	-2.232E+00	0.272	0.004	0.084
	(2.791E+00)	(4.256E-02)	(8.373E-01)			
	8.214E+00	1.049E+02	-1.753E+01	0.333	0.046	0.098
	(2.738E+00)	(4.857E+00)	(1.033E+01)			

FIGURE 12-8.
Typical output from *FFT-NLLS*.

RAE can serve as a measure of assessing the significance of each cosine component determined by the *FFT-NLLS* procedure. In this specific example, the period with the smallest RAE of 0.078 is that with a period of 20 ± 0.028, amplitude 35.74 ± 2.773, and phase 7.95 ± 0.2726. There are three other periodic components identified, however, with comparable RAE, and further analyses are needed in order to decide what role these additional periodic components play in the rhythmic behavior of the event.

A point of note is that the time of Phase(max) is actually the negation of the phase value as reported in the output created by FFT-NLLS (because of a historical convention in the circadian biology research community). As such, the actual time of Phase(max) in this example is at -7.95 hours relative to time zero in the data being analyzed.

FFT-NLLS is specifically designed to process data sets that are relatively short and/or noisy and is generally capable of extracting relatively weak rhythms. In addition, it extracts meaningful periods despite mean and variance nonstationarities that may exist in the data. It assigns period, phase, amplitude, and rhythmic strength/level of rhythmic determination (via the RAE) in a totally automated manner that is free of user-introduced analytical bias.

Sometimes, however, it is important to obtain a dynamic characterization of the rhythm, especially if it is important to know whether the period remains stable with time, stretches, or compresses [as in Figures 12-6(B) and 12-7(C)]. The algorithm *PHASEREF* that we describe next can be used in such cases.

PHASEREF. This method has been developed for assessing period, oscillatory amplitude, and phase information from a rhythmic data series. *PHASEREF* is a maximally assumption-free strategy in which no model form for any rhythms is assumed. However, interpretation of results does require the assumption that there exists in the data series being analyzed one dominant, primary rhythmic component, the period of which is (approximately) known a priori. *PHASEREF* is a modification of a method presented by Meerlo *et al.* (1997). *PHASEREF* requires the user to provide two filter windows with which to calculate two sets of smoothed running average values of the data series to be analyzed.

One smoothing filter is to possess a period value close to that expected for the dominant rhythm being expressed in the data series (i.e., for the case of circadian time series, this value would be approximately 24 hours). The result of smoothing the data with this period filter is to produce an approximately perfectly smoothed baseline series in which the dominant-period rhythm is effectively (nearly) completely removed [only long-term trend(s) may remain in the resultant smoothed series; that is, trend(s) that are characteristic of spectral components that possess periods longer than the 24-hour filter period]. A second, shorter-period filter length is also specified, the purpose of which is to generate a smoothed data series in which high-frequency noise is filtered out, but in which is retained the presumed longer-period dominant rhythm (i.e., the approximately 24-hour rhythm, in the case of circadian time series), as well as any longer-period trend(s). The difference series is then calculated between these two smoothed series (specifically, the short-period smoothed series minus the long-period smoothed series), from which are extracted the times of occurrence of four phase-reference conditions—up-cross, maximum, down-cross, and minimum (see Figure 12-9). In the outputs presented in the last section of this project, we have used *PHASEREF* with values of 24 hours and 12 hours for the filters.

Period estimates for the dominant rhythm are estimated from successive differences between up-crosses, maxima, down-crosses, and minima,

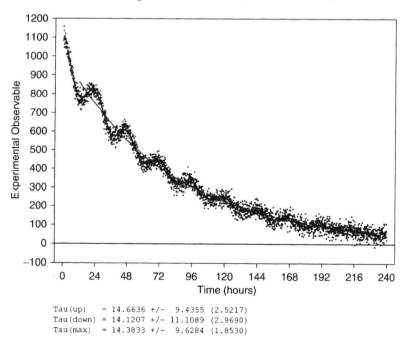

```
Tau(up)   = 14.6636 +/-  9.4355 (2.5217)
Tau(down) = 14.1207 +/- 11.1089 (2.9690)
Tau(max)  = 14.3833 +/-  9.6284 (1.8530)
```

FIGURE 12-9.
PHASEREF analysis of a noisy, nonstationary cosine wave, calculated with 6-hour and 24-hour filters. This figure shows sufficient noise is present to confound a clean assessment of period, arising primarily at the long-time end of the data series (the details of which are presented in Figure 12-10).

respectively (thus effectively providing for four mechanisms for estimating the dominant rhythmic periodicity). Rhythmic amplitude can be assessed by values produced for maxima and minima. Rhythmic phase is assessable in terms of the times of occurrence of maxima (assuming that acrophase is to be used as the phase reference marker of record). Thus, a typical implementation of this analytical strategy might entail calculating a 6-hour and a 24-hour running average of the original data series. The crossings of these two smoothed lines provide the rising and falling phase markers for each cycle. The maximum differences between the smoothed curves for each cycle (i.e., the peak and the trough) allow calculating the amplitude of each cycle. Time of peak provides a third phase marker. More details can be found in Abe *et al.* (2002).

The actual verbose output of the *PHASEREF* algorithm is quite technical and can be rather intimidating to a person without specific training in using the algorithm. To illustrate this statement (and only for this purpose!), we present, in Figure 12-10, the verbose output for the data depicted in Figure 12-9. In what follows, we shall not work with this raw output generated by *PHASEREF*. Instead, we shall present the relevant information in phase change plots such as those presented in the bottom panels of Figure 12-7.

The presence of noise may prove a challenge for the *PHASEREF* algorithms, as sufficient noise may make the calculation of clean estimates for the times of up-crossings, maxima, down-crossings, and

XOVER(up)	XOVER(down)	AMPL(max)	TIME(max)	ALPHA	RHO	TAU(up)	TAU(down)	TAU(max)
1.801500E+01								
	3.003500E+01	6.147595E+01	2.475000E+01	1.202000E+01				
4.241500E+01		-5.964130E+01	3.660000E+01		1.238000E+01	2.440000E+01		
	5.350500E+01	5.369855E+01	4.765000E+01	1.109000E+01			2.347000E+01	2.290000E+01
6.593500E+01		-4.243335E+01	6.015000E+01		1.243000E+01	2.352000E+01		2.355000E+01
	7.809500E+01	2.992264E+01	7.305000E+01	1.216000E+01			2.459000E+01	2.540000E+01
8.997500E+01		-3.398944E+01	8.305000E+01		1.188000E+01	2.404000E+01		2.290000E+01
	1.013450E+02	2.957397E+01	9.535000E+01	1.137000E+01			2.325000E+01	2.230000E+01
1.013550E+02		-1.528931E-02	1.013500E+02		1.000214E-01	1.138000E+01		1.830000E+01
	1.015650E+02	4.588318E-01	1.014500E+02	2.099991E-01			2.200012E-01	6.099998E+00
1.145950E+02		-2.159099E+00	1.069500E+02		1.303000E+01	1.324000E+01		5.599998E+00
	1.256750E+02	2.053612E+01	1.213500E+02	1.108000E+01			2.411000E+01	1.990000E+01
1.388550E+02		-1.934850E+01	1.306500E+02		1.317999E+01	2.425999E+01		2.370000E+01
	1.498650E+02	1.715137E+01	1.452500E+02	1.101001E+01			2.419000E+01	2.390000E+01
1.611250E+02		-1.900436E+01	1.543500E+02		1.125999E+01	2.227000E+01		2.370001E+01
	1.723950E+02	1.754298E+01	1.657500E+02	1.127000E+01			2.253000E+01	2.050000E+01
1.862750E+02		-1.420439E+01	1.797500E+02		1.387999E+01	2.514999E+01		2.539999E+01
	1.974150E+02	1.120005E+01	1.905500E+02	1.114000E+01			2.501999E+01	2.480000E+01
1.983450E+02		-2.171257E+00	1.980500E+02		9.300079E+00	1.207001E+01		1.830000E+01
	1.984350E+02	3.676605E-02	1.983500E+02	8.999634E-02			1.020004E+00	7.800003E+00
2.098350E+02		-7.790092E+00	2.008500E+02		1.140001E+01	1.149001E+01		2.800003E+00
	2.171150E+02	9.428452E+00	2.135500E+02	7.279999E+00			1.868001E+01	1.520000E+01
2.216850E+02		-5.838020E+00	2.195500E+02		4.569992E+00	1.184999E+01		1.870000E+01
	2.218350E+02	4.202080E-01	2.217500E+02	1.500092E-01			4.720001E+00	8.199997E+00
2.219850E+02		-6.716156E-02	2.218500E+02		1.499939E+01	3.000031E+01		2.300003E+00
	2.220550E+02	4.871750E-02	2.220500E+02	6.999207E-02			2.199860E-01	3.000031E-01
2.222750E+02		-6.159210E-01	2.221500E+02		2.200012E-01	2.899933E-01		2.999878E-01
	2.225150E+02	5.733643E-01	2.223500E+02	2.400055E-01			4.600067E-01	3.000031E-01
2.233050E+02		-1.382374E+00	2.231500E+02		7.899933E-01	1.029999E+00		1.000000E+00
	2.277250E+02	4.844704E+00	2.265500E+02	4.420013E+00			5.210007E+00	4.199997E+00

FIGURE 12-10.
Verbose, tabulated summary of the *PHASEREF* analytical session presented in Figure 12-9. Whereas the expected estimates for TAU(up), TAU(down), and TAU(max) are to be in the vicinity of 24 hours, numerous instances in which values for period estimates are considerably shorter than 24 hours arise. This is a consequence of noise confounds creeping in, beginning at about 100 hours of x-axis time, and manifesting consistently beyond about 200 hours of x-axis time.

minima difficult. In such cases, preprocessing the data through a filter to remove certain high frequencies may be beneficial to facilitate the analyses.

ARFILTER. Data filtering is generally applied to remedy the presence of noise. The data may be confounded either with additive noise (a situation that is readily dealt with) or with mixed noise (not often encountered, and fortunately so, because this type of noise confound is difficult to deal with). As described earlier, data filtering may either lead to information loss (so-called leaky filters) or to altering the data by introducing, for example, a phase change in the data. The filtering algorithm used for the examples in this chapter is called *ARFILTER.* Its implementation uses a forward–backward linear exponential (i.e., first-order) autoregressive filtering strategy, as reported in Orr and Hoffman (1974) (as apparently partially, at least, derived from acknowledged mention in a set of lecture notes at Princeton by Tukey in 1963), and we refer the reader to this article for the mathematical description and details. It should be stressed, however, that a particularly attractive feature of *ARFILTER* is that it results in *zero phase change* of the output.

If the input time series for *ARFILTER* is $x(t)$, the algorithm produces an output series $u(t)$ that represents the low-frequency component(s) of the original data series. The filtering procedure uses a single parameter ρ, the value of which is optimized such that the residuals, defined by $[x(t) - u(t)]$, are least-squares minimized. The detrended residuals $[x(t) - u(t)]$ then represent the residual high-frequency component(s) that were filtered out by this forward–backward autoregressive filtering process. It is important to note that, depending on the situation, either $u(t)$ or $[x(t) - u(t)]$ may be used in subsequent analysis: $u(t)$ if the objective was to remove excessive high-frequency noise from the original data series and $[x(t) - u(t)]$ if the objective was to remove confounding, low-frequency trending from the original data series. Figure 12-11 demonstrates the effect of applying *ARFILTER* on the noisy, nonstationary cosine wave from Figure 12-9. The time series $u(t)$ and $[x(t) - u(t)]$ are presented to the right of the original time series $x(t)$. As with any noise removal algorithm, an important question to always consider is whether the rhythmic characteristics of the *ARFILTER*ed data remain the same as, or at least reasonably close to, those present in the original unfiltered data.

DTRNDANL. The removal of a slow, gradual change (or drift) from the time series is called detrending. The presence of a trend reflects the change in some quantity or property (e.g., as we already noted, gradual depletion of luciferin causing a decline in the average levels of bioluminescence). Detrending may, in fact, be viewed as a special type of filtering for which the goal is to remove the lowest frequencies. The residual data are then used for analyses, because for these data the low frequencies (that is, the trend) have been removed.

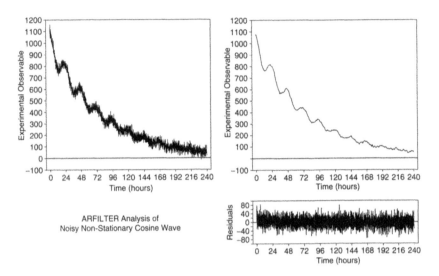

FIGURE 12-11.
Example of *ARFILTER* applied to a noisy, nonstationary cosine wave.

DTRNDANL applies a detrending algorithm in which the user specifies the following inputs:

1. The value of a filter window (FWINDOW) to apply; and

2. Whether the detrended data are to be presented in original y-value-space or in terms of standard normal deviates (SND-space), with the latter being used to address variance nonstationarites. This option is useful when the data present variance nonstationarities, because presenting the data on a scale of standard normal deviates alleviates the problem.

The selection of an appropriate value for a filter window is critical for successful application of this algorithm. For example, to apply *DTRNDANL* to circadian rhythms data that are recorded in units of hours, a value for FWINDOW of 24 hours would be an appropriate choice (assuming that the dominant rhythm exhibited by the data is near 24 hours in period). Figure 12-12 visualizes the output of *DTRNDANAL* with input provided by the data sets from Figure 12-9.

We are now ready to consider the following problem.

DETERMINING THE EFFECTS OF A TRANSLATIONAL INHIBITOR ON THE CIRCADIAN CLOCK

The file *PRE.NO* (from www.biomath.sbc.edu/data.html) is simulated to represent 10 days of in vitro (therefore, free-running) *per-luc* activity before a 6-hour treatment with a translational inhibitor. The sample was isolated from an animal that had been exposed long-term (for at least a couple of weeks) to a 12-hour light:12-hour dark cycle.

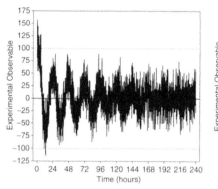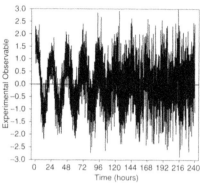

FIGURE 12-12.
An example of *DTRNDANL* applied to the noisy, nonstationary cosine wave originally presented in Figure 12-9. The panel on the left is the result produced by leaving the data series being detrended in original data space, whereas the panel on the right is the result produced by processing and converting the data series being detrended to standard normal deviate space (SND-space).

The file *POST.NO* (also from www.biomath.sbc.edu/data.html) is simulated to represent the 10 days of *in vitro* (free-running) *per-luc* activity observed immediately after washing out the translational inhibitor after six hours of treatment.

The primary question being addressed regards what effect the treatment with translational inhibitor had on the functioning of the circadian clock and/or the observed *per-luc* expression patterning.

The following list presents just a few supposed scenarios that could occur upon the removal of the translational inhibitor from the cells:

1. The translational inhibitor has destroyed the circadian clock of the cells.

2. The circadian clock has been suppressed or delayed during the six-hour treatment with the translational inhibitor, but not destroyed. Upon removal of the translational inhibitor, the circadian clock resumes its pretreatment rhythms, picking up from where it has left off when the inhibitor was applied.

3. The circadian clock has been suppressed or delayed during the 6-hour treatment with the translational inhibitor, but not destroyed. Upon removal of the translational inhibitor, the circadian clock resumes a rhythm, but the rhythm has different circadian characteristics compared to the pretreatment rhythm.

4. The circadian clock has been neither destroyed nor completely suspended. It has continued to function (perhaps with an altered rhythm) during the 6-hour treatment with translational inhibitor and the function continues after the inhibitor is removed.

Your goal for this project is to quantitatively characterize, as thoroughly as possible, the circadian rhythms from both the *PRE.NO* and *POST.NO* data, and use these characterizations to determine which of the above scenarios (or any of the number of alternative hypothetical scenarios that you will formulate) is most likely to have occurred.

Note that this is an open-ended project with a complexity level much higher compared with the previous laboratory projects in the Manual. We encourage you to play with the data, hypothesize and explore, and aim at finding the best results possible given the limited data available. We hope that this project will provide you with a first-hand experience of what doing research in the field of circadian rhythms is about and that you will find the experience rewarding.

To facilitate your work, we have provided the outputs from applying the algorithms, *DTRNDANL*, *FFT-NLLS*, and *PHASEREF* to the *PRE.NO* and *POST.NO* data. Outputs of *FFT-NLLS* and *PHASEREF* are also presented for the detrended data (that is, the data preprocessed with *DTRNDANL*) and for the noise filtered data (that is, the data preprocessed with *ARFILTER*). You may find some of the outputs useful and some irrelevant to the questions that you would want to answer. It will be up to you to decide which particular analyses to use and how to combine the use of various analyses.

Exercise 12-1

Plot the data files *PRE.NO* and *POST.NO*. You may use *MS Excel*, *MINITAB*, or any other software of your choice to do that. Observe the graphs and describe each one in a paragraph, giving your overall impressions as well as any specific characterizations you draw from each graph.

Exercise 12-2

Using the analyses of the data provided (or any other analyses of your choice that you can perform on the data), state your hypothesis on the effect that the translational inhibitor has had on the function of the circadian clock of the cells.

Exercise 12-3

Use the results from the data analyses to quantitatively corroborate your hypothesis as thoroughly as possible.

Exercise 12-4

Present a summary of your conclusions.

Exercise 12-5

Are there any additional analyses that you wish we had performed, in addition to those whose outputs we have presented here, in order to allow you to more fully corroborate your hypothesis? If so, list them, and give your rationale.

Exercise 12-6

Suggest follow-up experiments that will allow for a further validation of your conclusions.

SOFTWARE OUTPUTS

The following Figures 12-13 through 12-18 provide outputs from *FFT-NLLS* and *PHASEREF* applied to the original data or data preprocessed with *ARFILTER* and/or *DTRNDANL*.

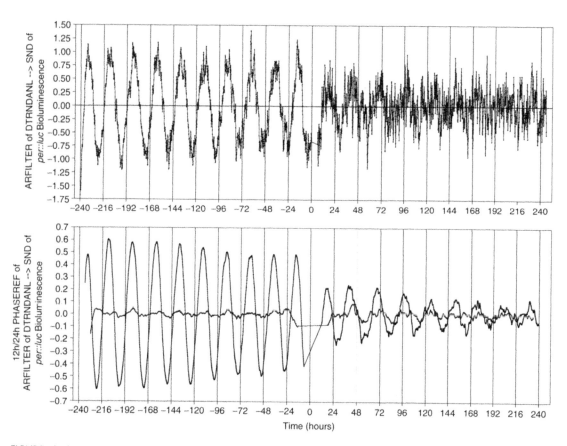

FIGURE 12-13.
The top panel presents the detrended data after noise-reduction filtering using *ARFILTER*. The resulting data series depicted in this panel thus correspond to detrended, normalized data that were subsequently filtered to reduce high-frequency noise. The bottom panel depicts the results of applying the *PHASEREF* algorithm to the detrended, noise-reduced data series from the top panel in which *PHASEREF* was applied with a combination of a 12-hour filter (black line) and a 24-hour filter (gray line).

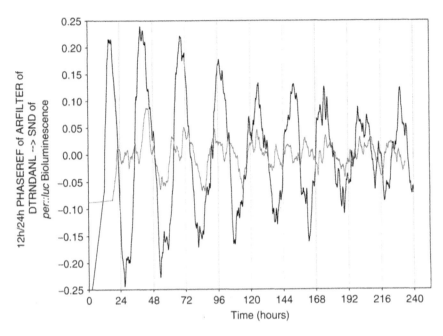

FIGURE 12-14.
This expanded view of the post-stimulus data series of the previous bottom panel (i.e., the data series occurring during positive time on the x-axis scale) makes apparent the difficulties in assigning unambiguous crossover times for some of the crossovers in the vicinity of noisy 12-hour (black) and 24-hour (gray) smoothed curves.

```
VARIANCE OF FIT           =     2.0214E-02 WITH    6 DEGREES OF FREEDOM
SUM OF RESIDUALS SQUARED  =     1.2128E-01
SQUARE ROOT OF VARIANCE   =     1.4218E-01

FITTED PARAMETER VALUES AND CONFIDENCE LIMITS
DC   =  1.095016E+01 +/-  4.155998E-01 ( 1.053456E+01 ->  1.136576E+01)
Slop =  4.944612E-01 +/-  6.975701E-02 ( 4.247042E-01 ->  5.642182E-01)
```

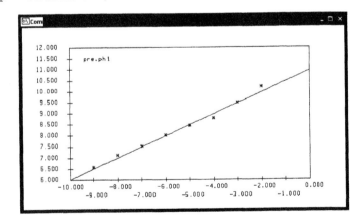

FIGURE 12-15.
Plots of the daily times of maxima for the *PRE.NO* data obtained by using *PHASEREF* with 12-hour/24-hour filters. For the prestimulus data, this panel depicts, on the y-axis, the daily times of phase maxima occurring modulo-24-hours (i.e., reported within the context of each 24-hour day), and, on the x-axis, the day number preceding time zero. The straight line is a least-squares estimate of the best-fit linear regression line to these data. The parameter *DC* (10.95 ± 0.42) is the y-intercept value on day 0 (i.e., the time of phase maximum expected on day 0). The parameter Slop (0.494 ± 0.070) is the slope of the linear regression line in units of hours/day (the interpretation of which is that the time of phase maximum is occurring 0.494 ± 0.070 hours later each consecutive day during the prestimulus period during which data were acquired.

```
VARIANCE OF FIT          =      2.3317E+00 WITH 5 DEGREES OFFREEDOM
SUM OF RESIDUALS SQUARED =      1.1658E+01
SQUARE ROOT OF VARIANCE  =      1.5270E+00

FITTED PARAMETER VALUES AND CONFIDENCE LIMITS
DC   = 1.705757E+01 +/-   4.306780E+00  ( 1.275079E+01 -> 2.136435E+01)
Slop = 2.508047E+00 +/-   9.630280E-01  ( 1.545019E+00 -> 3.471075E+00)
```

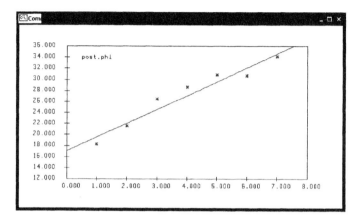

FIGURE 12-16.

Plots of the daily times of maxima for the *POST.NO* data obtained by using *PHASEREF* with 12-hour/24-hour filters. For the post-stimulus data, this panel depicts, on the y-axis, the daily times of phase maxima occurring modulo-24-hours (i.e., reported within the context of each 24-hour day), and, on the x-axis, the day number following time zero. The straight line is a least-squares estimate of the best-fit linear regression line to these data. The parameter *DC* (17.1 ± 4.3) is the y-intercept value on day 0 (i.e., the time of phase maximum expected on day 0). The parameter *Slop* (2.51 ± 0.96) is the slope of the linear regression line in units of hours/day (the interpretation of which is that the time of phase maximum is occurring 2.51 ± 0.96 hours later each consecutive day during the post-stimulus period during which data were acquired. However, there appears to be some rather marked curvature in the actual data points that is not well characterized by the best-fit linear regression line.

FFT-NLLS of PRE.NO and POST.NO

File Name	Amplitude	Period	Phase	Rel-Amp	Rel-Per	Rel-Phi
pre.no	1.027E+04	3.957E+03	-1.855E+03	0.219	0.110	0.055
	(2.253E+03)	(4.361E+02)	(2.181E+02)			
	1.288E+02	2.450E+01	-1.099E+01	0.049	0.001	0.008
	(6.339E+00)	(3.025E-02)	(1.930E-01)			
post.no	2.286E+03	2.536E+03	1.144E+03	0.789	0.382	0.191
	(1.803E+03)	(9.690E+02)	(4.845E+02)			
	4.152E+01	2.731E+01	1.318E+01	0.266	0.011	0.042
	(1.105E+01)	(2.992E-01)	(1.150E+00)			

FIGURE 12-17.

FFT-NLLS performed on the original data.

FFT-NLLS of PRE.DTR and POST.DTR

File Name	Amplitude	Period	Phase	Rel-Amp	Rel-Per	Rel-Phi
pre.dtr	9.225E-01	2.429E+01	-9.913E+00	0.059	0.003	0.020
	(5.425E-02)	(7.916E-02)	(4.836E-01)			
post.dtr	2.238E-01	2.661E+01	9.937E+00	0.472	0.022	0.116
	(1.057E-01)	(5.934E-01)	(3.091E+00)			
	1.172E-01	2.411E-01	2.531E-02	0.902	0.000	0.238
	(1.057E-01)	(9.937E-05)	(5.744E-02)			

FIGURE 12-18.
FFT-NLLS performed on the detrended data.

REFERENCES

Abe, M., Herzog, E. D., Yamazaki, S., Straume, M., Tei, H., Sakaki, Y., Menaker, M., & Block, G. D. (2002). Circadian rhythms in isolated brain regions. *Journal of Neuroscience, 22*, 350–356.

Meerlo, P., van den Hoofdakker, R. H., Koolhaus, J. M., & Daan, S. (1997). Stress-induced changes in circadian rhythms of body temperature and activity in rats are not caused by pacemaker changes. *Journal of Biological Rhythms, 12*, 80–92.

Orr, W. C., & Hoffman, H. J. (1974). A 90-min cardiac biorhythm: Methodology and data analysis using modified periodograms and complex demodulation. *IEEE Transactions on Biomedical Engineering, 21*, 130–143.

Plautz, J. D., Straume, M., Stanewsky, R., Jamison, C. F., Brandes, C., Dowse, H. B., Hall, J. C., & Kay, S. A. (1997). Quantitative analysis of *Drosophila period* gene transcription in living animals. *Journal of Biological Rhythms, 12*, 204–217.

Tukey, J. W. An introduction to the frequency analysis of time series. In Thompson, J.R., & Brillinger, D.R. (eds.). *Mathematics 596*. Princeton, NJ: Princeton University.

FURTHER READING

Matsuno, H., Tanaka, Y., Aoshima, H., Doi, A., Matsui, M., & Miyano, S. (2003). Biopathways representation and simulation on hybrid functional Petri net. *Silico Biology, 3*, 32.

Printed and bound by CPI Group (UK) Ltd, Croydon, CR0 4YY

03/10/2024

01040315-0015